Cover Image Credit/Copyright Attribution: IIlerlok_Xolms/Shutterstock

Bálint Németh · Lambros Ekonomou
Editors

Flexitranstore

Special Session in the 21st International Symposium on High Voltage Engineering (ISH 2019)

Editors
Bálint Németh
Head of High Voltage Laboratory
Budapest University of Technology
and Economics
Budapest, Hungary

Lambros Ekonomou
Department of Electrical and Electronic
Engineering Educators
ASPETE-School of Pedagogical
and Technological Education
Athens, Greece

https://doi.org/10.1007/978-3-030-37818-9

Contents

Assessing the Operational Flexibility in Power Systems with Energy Storage Integration

Lysandros Tziovani[1(✉)], Maria Savva[1], Markos Asprou[1], Panayiotis Kolios[1], Elias Kyriakides[1], Rogiros Tapakis[2], Michalis Michael[2], and Christos Hadjilaou[2]

[1] KIOS Research and Innovation Center of Excellence and Department of Electrical and Computer Engineering, University of Cyprus, Nicosia, Cyprus
ltziov01@ucy.ac.cy

[2] Transmission System Operator Cyprus, Nicosia, Cyprus

Abstract. Increased level of flexibility is essential in power systems with high penetration of renewable energy sources in order to maintain the balance between the demand and generation. Actually, the flexibility provided by energy storage systems and flexible conventional resources (i.e., generating units) can play a vital role in the compensation of the renewable energy sources variability. In this paper, the flexibility of the conventional generating units is quantified and incorporated in a unit commitment model in order to evaluate the impact of different system flexibility levels on the optimal generation dispatch and on the operational cost of the power system. An emerging flexible option such as the battery storage is included in the unit commitment formulation, evaluating the flexibility contribution of the storage and its effect on the system operational cost. In this paper, the flexibility of a real power system is assessed while the unit commitment problem is formulated as a mixed-integer linear program. The results show that the integration of a storage unit in the power generation portfolio provides a significant amount of flexibility and reduces the system operational cost due to the peak shaving and valley filling.

Keywords: Energy storage · Operational flexibility · Unit commitment

1 Introduction

The increasing integration of the renewable energy sources (RES) into the power system causes several operating and security problems due to the inherent stochastic nature of RES. A problem that might arise due to the stochasticity of the power generation from RES is the maintenance of the power balance in case that the share of renewables in the generation mix is very large. This means that sufficient resources must be scheduled by system operators in order to maintain the load-generation balance at all instances. In other words, the operators should have the flexibility to response in any abrupt changes of the power generation from RES [1, 2]. Flexibility can improve the efficient power system operation, and as a result to reduce the operating cost, consumer prices and the operational risk. The flexibility can be provided by flexible resources such as generating units, energy storage systems (ESSs), interconnections

B. Németh and L. Ekonomou (Eds.): ISH 2019, LNEE 610, pp. 1–12, 2020.
https://doi.org/10.1007/978-3-030-37818-9_1

with neighboring systems, demand response schemes, grid strength and forecasting accuracy [3, 4].

The flexibility provided by conventional generating units is very important since these units are the main resource for satisfying the net load. The flexibility potential of the generating units depends by their present operational state and their technical characteristics such as operational range, ramping rates, start-up time, shut-down time, and the minimum up and down times. There are several indices in the literature to assess the flexibility of the system. In [5–7], a flexibility index is defined in order to quantify the flexibility of each type of generating unit. This index is mainly based on the technical characteristics of each unit. Also, the calculated flexibility index in [6] is incorporated in a unit commitment (UC) model to investigate the impact of generators flexibility in the day-ahead operational decisions. In [7], the flexibility indices of the generators are integrated in a dedicated unit construction and commitment (UCC) algorithm in order to determine the optimal investments in flexible generating units. Results show that the increase of the system flexibility can handle more uncertainties in load and generation. At the same time though the greater flexibility provided by conventional generating units results in an additional operating cost.

ESSs are an emerging technology that can reduce the operational cost of the system in valley filling and peak shaving applications, due to their ability to absorb power during low-cost hours and to inject power during high-cost hours [8, 9]. Also, ESSs are very flexible resources that can be used in compensating the intermittent and uncontrollable character of RES. In [10], an ESS is incorporated in a UC model in order to evaluate the storage profits under a high penetration of RES. Similarly, in [11], the integration of ESSs in a stochastic UC model reduces the wind power curtailment and the system operation cost under a high penetration of wind power.

This paper evaluates the operational flexibility provided by conventional generating units in a real isolated power system. The flexibility of the system is quantified according to [5], and the resulting flexibility indices are incorporated in a UC model as a constraint, in order to assess the impact of different system flexibility levels on the optimal generation dispatch and on the operational cost of the system. Further, the flexibility contribution of a possible integration of a battery storage in the system, and its effect on the system operational cost is investigated. The UC problem of the system with the integration of the battery storage and the consideration of the flexibility index is formulated as a mixed-integer linear program which is solved by a commercial solver. Finally, the total upward and downward flexibility of the system is calculated through the generation dispatch and the storage operation of the UC solution for different levels of system flexibility provided by the conventional units. The main contributions of this work are the following:

(a) The proposed formulation for the flexibility assessment is applied to a real power system in which the impact of the integration of a battery storage on the flexibility and the operating cost of the system are examined.
(b) In [12], the total upward and downward flexibility provided by the conventional units and the battery storage is determined without the incorporation of the units' flexibility indices into the UC. In this work, the units' flexibility indices are incorporated in the UC, and the total upward and downward flexibility is

calculated for different minimum levels of system flexibility provided by the conventional units.

The rest of this paper is organized as follows. In Sect. 2, the flexibility indices of the generating units are calculated. Also, in Sect. 3, the UC formulation is presented, while the calculation of the upward and downward flexibility is following in Sect. 4. Simulation results are presented in Sect. 5 and the paper concludes in Sect. 6.

2 Flexibility Index Calculation

The flexibility index of each generating unit is determined by constructing the composite flexibility metric (CFM) proposed in [5]. The flexibility index is calculated based on seven technical characteristics of the generating units namely, the minimum stable generation level, operating range, ramp up/down capabilities, start-up time, and minimum up and down time. Since the technical characteristics are expressed in different units of measurements and in disproportionate scales a normalization is required. The normalization procedure ensures that an increase in the value of a technical characteristic increases also the flexibility index. Moreover, the normalized characteristics can be compared and aggregated. The min-max normalization technique is used to convert all the technical characteristics to a similar range between 0 and 1 as shown in (1).

$$I_{ji} = \frac{x_{ji} - min_i(x_{ji})}{max_i(x_{ji}) - min_i(x_{ji})} \tag{1}$$

where x_{ji} is the value of technical characteristic j of unit i, I_{ji} is the normalized value of x_{ji}, while $min_i(x_{ji})$ and $max_i(x_{ji})$ are associated with the minimum and maximum values respectively of the technical characteristic j across all the units.

The next step for the construction of the flexibility index is the weighting. The weighting model of [5] is used for assigning weight to each technical characteristic, which reflect the importance of each characteristic in the provision of flexibility. According to the weighting model, the most important characteristics of the units in the provision of flexibility are the minimum stable generation level, the operating range, and the start-up time. In the last step of the CFM calculation, the flexibility index ($Flex_i$) of each unit is calculated using (2), by calculating the summation of the normalized values of the technical characteristics (I_{ji}) multiplied by their associate weights (w_j).

$$Flex_i = \sum_{j=1}^{K} I_{ji} \times w_j + \sum_{j=1}^{J} (1 - I_{ji}) \times w_j \tag{2}$$

where K and J are the technical characteristics which are positively and negatively correlated with the supply of flexibility respectively. It must be noted that the value of the flexibility index of each generating unit ($Flex_i$) is between 0 and 1, and the highest value determine the most flexible unit.

3 Unit Commitment Formulation

In this section, a UC formulation is proposed which incorporates the calculated flexibility indices, in order to investigate the impact of different flexibility levels on the optimal generation dispatch and on the operational cost of the system. Also, a battery storage is included in the unit commitment formulation, in order to evaluate the flexibility contribution of the storage and its effect on the system operational cost. The unit commitment problem is formulated as a mixed-integer linear program (MILP) along an arbitrary time horizon T with one hour time intervals. The quadratic cost function of the generating units is approximated by a piece-wise linear function in order to use the MILP technique to solve the problem. Note that the accuracy of this approximation can be controlled by the number of the linear segments considered.

3.1 Objective Function

The objective function of this optimization problem is presented in (3) and targets to minimize the total operational cost of the system. More specifically, the generation and the commitment cost of the generating units is minimized for the whole period of study. In this work, zero marginal cost is assumed for RES power penetration and a strictly priority dispatch is maintained. Also, it is assumed that no RES generation curtailments as well as no load shedding is performed.

$$\min f(cost) = \sum_{t=1}^{T} \sum_{i=1}^{U} \left(F_i\left(p_i^t\right) + b_i^t N_i + z_i^t S_i \right) \tag{3}$$

where T and U are the time periods (hours) and the number of generating units; variables p_i^t, b_i^t and z_i^t are the generation (continues), on/off states (binary) and the start-up state (binary) of unit i at time t; $F_i\left(p_i^t\right)$ is the piece-wise linear cost function of the generation p_i^t for unit i at time t; N_i and S_i are the no-load cost and the start-up cost of unit i.

3.2 Constraints

The following equations presents all the constraints and conditions of the model:

1. *System Flexibility*: The total system flexibility of the power system is calculated as the summation of the flexibility indices ($Flex_i$) of the committed units for the whole simulation period and must be greater than the minimum total system flexibility (MTF). A larger value of MTF forces more flexible units to be committed in order to satisfy the system flexibility constraint which is shown in (4). Note that $Flex_i$ and MTF are expressed in per-unit (p.u).

$$\sum_{t=1}^{T} \sum_{i=1}^{U} Flex_i \times b_i^t \geq MTF \tag{4}$$

2. *Generation Limits & Units Start-up*: The produced power from each committed unit must be within a range defined by their minimum (P_i^{min}) and maximum (P_i^{max}) operating limits. Also, the produced power from the decommitted units must be zero. As a result, p_i^t is a discontinued variable and this constraint is shown in (5). The constraint in (6) forces the start-up variables (z_i^t) to take the value 1 when the corresponding units were off at time $t-1$ and switch on at time t. Thus, the start-up cost of these units is added to the objective function. In every other case, these variables are set to zero.

$$\begin{aligned} p_i^t &\leq P_i^{max} \times b_i^t, \quad \forall i, \forall t \\ p_i^t &\geq P_i^{min} \times b_i^t, \quad \forall i, \forall t \end{aligned} \tag{5}$$

$$-b_i^{t-1} + b_i^t \leq z_i^t, \quad \forall i, \forall t \in [2, T] \tag{6}$$

3. *Power Balance*: The produced power from the committed units plus the discharging power (p_{Dis}^t) of the battery must be equal to the net load (D^t) plus the charging power (p_{Ch}^t) of the battery for each time interval t. D^t is defined as the load demand minus the RES power. s_d and s_c are the discharging and charging coefficients of the battery.

$$\sum_{i=1}^{U} p_i^t + s_d \times p_{Dis}^t = \frac{1}{s_c} \times p_{Ch}^t + D^t, \quad \forall t \tag{7}$$

4. *Spinning Reserve*: The available power from the committed units must be greater than summation of D^t and the spinning reserve (R^t) excluding the discharging power of the battery. In this work, the battery storage does not contribute in the supply of reserve, however the discharging power of the battery is subtracted from the net load.

$$\sum_{i=1}^{U} b_i^t \times P_i^{max} \geq D^t + R^t - s_d \times p_{Dis}^t, \quad \forall t \tag{8}$$

5. *Primary Reserve*: The available primary reserve from the committed units must be equal or greater than the minimum primary reserve $(PR^{min,t})$ of the system for each time interval t (9). The available primary reserve (q_i^t) of each committed unit is defined as the difference between the maximum and the actual generation of the unit (10). Also, the available primary reserve of each committed unit must be less than its maximum (Q_i^{max}) bound (11).

$$\sum_{i=1}^{U} q_i^t \geq PR^{min,t}, \quad \forall t \tag{9}$$

$$q_i^t \leq P_i^{max} - p_i^t, \quad \forall i, \forall t \tag{10}$$

$$q_i^t \leq Q_i^{max} \times b_i^t, \qquad \forall i, \forall t \tag{11}$$

6. *Minimum Up & Down times*: There is a minimum time (Up_i) that units must remain online after their commitment (12). Also, there is a minimum time (Dp_i) that generating units must remain offline after their decommitment (13).

$$\begin{gathered} b_i^t - b_i^{t-1} \leq b_i^{T_{up}}, \\ \forall i, T_{up} \in [t+1, \min\{t+Up_i-1, T\}], \\ \forall t \in [2, T-1] \end{gathered} \tag{12}$$

$$\begin{gathered} b_i^{t-1} - b_i^t \leq 1 - b_i^{T_{down}}, \\ \forall i, T_{down} \in [t+1, \min\{t+Dp_i-1, T\}], \\ \forall t \in [2, T-1] \end{gathered} \tag{13}$$

7. *State of Charge*: The state of charge (SOC_t) of the battery storage in time t is measured in MWh and is expressed as the initial capacity (IC) of the storage minus the summation of the discharging power plus the summation of the charging power for all the past and the present time intervals. Also, SOC_t must be within the minimum (SOC_{min}) and maximum (SOC_{max}) state of charge of the battery storage in MWh. The state of charge constraint is presented in (14). Note that the energy of the battery (SOC) with the charging and discharging power of the battery are related since, in the simulations only one hour time intervals are considered.

$$\begin{gathered} SOC_t = IC - \sum_{\tau=1}^{t} p_{Dis}^{\tau} + \sum_{\tau=1}^{t} p_{Ch}^{\tau}, \quad \forall t \\ SOC_t \geq SOC_{min}, \; SOC_t \leq SOC_{max}, \; \forall t \end{gathered} \tag{14}$$

8. *Battery Storage Restriction*: The simultaneous charging and discharging of the battery storage at the same time interval is restricted through (15) and (16). Also, the charging and discharging power in MW must be less than the maximum charging (P_{Ch}^{max}) and discharging (P_{Dis}^{max}) power of the battery storage (15). Note that v_t and n_t are binary variables associated with the discharging and charging power respectively.

$$\begin{gathered} 0 \leq p_{Dis}^t \leq P_{Dis}^{max} \times v_t, \quad \forall t \\ 0 \leq p_{Ch}^t \leq P_{Ch}^{max} \times n_t, \quad \forall t \end{gathered} \tag{15}$$

$$v_t + n_t \leq 1, \quad \forall t \tag{16}$$

It should be noted that the ramp-up and ramp-down constraints of the generating units are ignored in this study, since the units can change their production from the minimum to the maximum and vice versa, in one-hour time interval (as considered in this study).

4 Upward and Downward Flexibility

The ability of the system to respond to abrupt power changes in a certain time period depends on the level of the flexibility that exist in the system. This capability of the system is defined as the upward and downward flexibility. In this section, the equations related to the total upward and downward flexibility provided by the generating units and the battery storage are shown. In this work, the calculation of the system upward and downward flexibility indicates the impact of the battery storage and the incorporated flexibility indices (to the unit commitment problem) on the ability of the system to respond to power changes.

The upward flexibility provided by the conventional generating units is presented in (17) and it is actually the extra power that can be produced by the committed units between the time intervals which are taken into account. In contrast, the downward flexibility (calculated as shown in (18)) is the total power that the committed units are able to reduce (based on the minimum stable operating level). In this work, based on the results of the UC the upward and downward flexibility of the system are calculated for the whole simulation period.

$$Units\ Flex_{Up} = \sum_{t=1}^{T} \sum_{i=1}^{U} b_i^t \times \left(P_i^{max} - p_i^t\right) \tag{17}$$

$$Units\ Flex_{Down} = \sum_{t=1}^{T} \sum_{i=1}^{U} b_i^t \times \left(p_i^t - P_i^{min}\right) \tag{18}$$

The provision of upward and downward flexibility from a battery storage depends on its current state of operation and is limited by its capacity and its current stored energy. For example, a battery storage is not able to provide upward flexibility if it is fully discharged and similarly, it cannot provide downward flexibility if it is fully charged. In (19) and (20) the upward and downward flexibility that can be provided by the battery in the time period T_1 where the battery operates in discharging mode is presented. Similarly, the upward and downward flexibility provided by the battery for the time periods (T_2 & T_3) where the battery operates at the charging and the non-working mode respectively can be calculated through (21)–(24). It is assumed that the battery storage cannot change its operation mode in the same time interval (i.e. change from charging to discharging mode). In reality though, the battery can change its state of operation in the same time interval providing greater amount of flexibility. This will be a part of the future work.

$$Flex_{Up}^{Dis} = \sum_{t=1}^{T_1} min\left[\left(P_{Dis}^{max} - p_{Dis}^t\right), SOC_t\right], T_1 \in T \tag{19}$$

$$Flex_{Down}^{Dis} = \sum_{t=1}^{T_1} p_{Dis}^t, \quad T_1 \in T \tag{20}$$

$$Flex_{Up}^{Ch} = \sum_{t=1}^{T_2} p_{Ch}^{t}, \quad T_2 \in T \tag{21}$$

$$Flex_{Down}^{Ch} = \sum_{t=1}^{T_2} min\left[\left(P_{Ch}^{max} - p_{Ch}^{t}\right), \left(SOC_{max} - SOC_t\right)\right], \quad T_2 \in T \tag{22}$$

$$Flex_{Up}^{NonWork.} = \sum_{t=1}^{T_3} min\left(P_{Dis}^{max}, SOC_t\right), \quad T_3 \in T \tag{23}$$

$$Flex_{Down}^{NonWork.} = \sum_{t=1}^{T_2} min\left(P_{Ch}^{max}, \left(SOC_{max} - SOC_t\right)\right), \quad T_3 \in T \tag{24}$$

In (19), the upward flexibility is defined as the minimum amount of energy between the stored energy (MWh), i.e. SOC and the power (MW) that the battery can provide (according to the maximum discharge of the battery). The flexibility in (20) and (21) can be provided by stopping the discharging or the charging of the battery, respectively. In (22), the downward flexibility is calculated as the minimum between the available extra charging power that the battery can absorb and its available capacity for storing energy. The upward flexibility of the non-working state (23) is defined as the minimum between the maximum discharging power and the state of charge of the battery, while in (24) the downward flexibility of this state is the minimum between the maximum charging power and the available energy that the battery can store.

5 Simulation Results

In this section the impact of the battery storage integration and the incorporation of the flexibility indices (in the unit commitment problem) on the system operating cost is investigated. The investigation is performed on the upward and downward flexibility of the isolated power system of Cyprus. The installed capacity of the conventional units of the power system of Cyprus is 1478 MW, while the installed capacity of wind and PV power is 157 MW and 125 MW respectively. The proposed MILP formulation is coded in Matlab and a commercial solver is used to solve the problem using real data of the power system of Cyprus. More specifically, this problem is solved for two consecutive summer days (48 h), using the load demand and the RES penetration of these days as input data. Those days, the maximum RES penetration reached 11.48% of the peak of the load. Note that a horizon of 24 h can also be applied in order to be associated with the day-ahead market, however a longer period of time is used in this work to examine the impact of the battery storage in a longer time period. Two batteries of 60 MWh and 120 MWh with maximum charging and discharging power of 60 MW and 120 MW are considered in the simulations [6]. It is assumed that the round trip efficiency of the batteries is 92%, and their aforementioned capacities are the usable capacities. Therefore, the nominal capacity of the batteries is greater.

The impact of the 120 MWh battery storage on the operation of the power system of Cyprus is presented in Fig. 1. More specifically, the generating power from the conventional units with and without the integration of the battery is illustrated in Fig. 1 (a) for the 48 h horizon. Also, the state of charge of the battery is presented in Fig. 1 (b). Initially, it is assumed that the battery is fully discharged. Note that the 120 MWh is the usable capacity of the battery, and we assume that there are no limitations about the minimum state of charge of the battery. The valley filling and peak shaving applications of the battery storage are obvious in Fig. 1(a). The integration of the battery storage manages to reduce the range between the maximum and the minimum produced power, with a great benefit on the operational cost of the system due to the fact that less generating units need to be committed and decommitted in order to satisfy the security and the technical constraints of the system.

Figure 2 illustrates the rise of the system operational cost in respect with the increment of the system flexibility (MTF) for the cases with and without the battery storage integration. Note that the system flexibility constraint become binding for values greater than 310 (p.u) for the case of the battery integration, and for values greater than 320 (p.u) for the case of no battery integration. This means that for smaller values, the system flexibility constraint does not affect the optimal solution. As it can be seen in this figure, the system operational cost increases significantly with the increase of the MTF due to the fact that more flexible generating units as well as additional generating units are needed to be committed in order to satisfy the flexibility

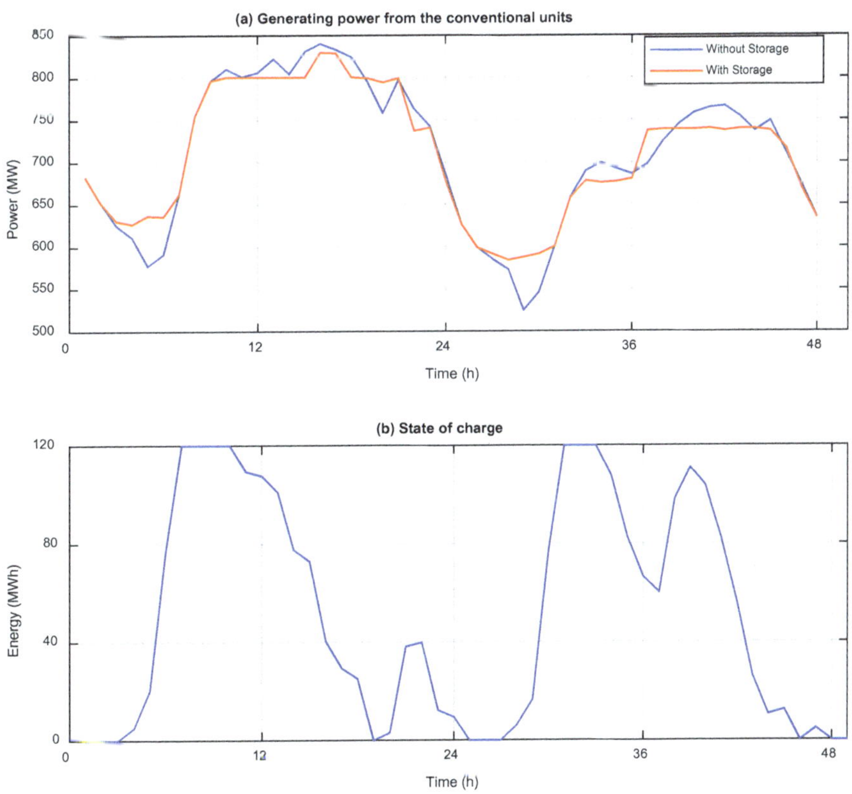

Fig. 1. Battery storage integration.

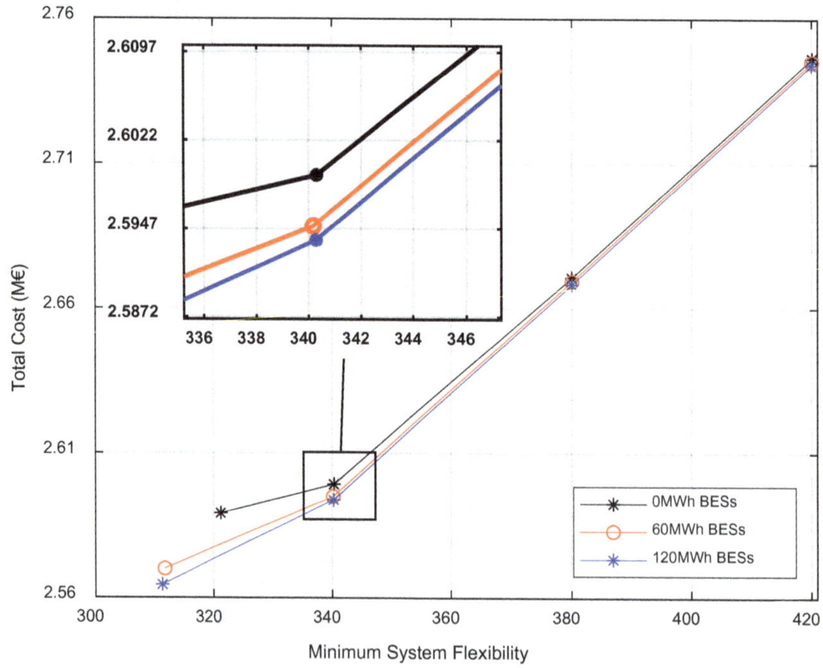

Fig. 2. Operational cost of the system.

requirements. Also, the cost reductions due to the integration of the battery storage are presented in Table 1 as the system profits compared to the case without the battery storage integration. The profit with the battery integration is very high when the system flexibility constraint is non-binding and is reduced significantly when this constraint becomes binding. This is mainly because only the generating units are considered as flexible resources in the current UC formulation. It should be noted that the consideration of the battery as a flexible resource will be the subject of future work.

The total upward and downward flexibility provided by the generating units and the battery storage is calculated for several values of the minimum flexibility and are presented in Tables 2 and 3. It is clear that the total upward flexibility increases dramatically as the minimum flexibility value increases, and this is due to the commitoment of additional generating units and their dispatch at lower operating levels. In contrast, the total downward flexibility decreases because the committed generating units are dispatched at lower operating levels. However, as the results show, this reduction on the downward flexibility is not very significant. The integration of a

Table 1. System profits with battery integration.

Minimum flexibility (p.u)	Profit 60 MWh BES (€)	Profit 120 MWh BES (€)
<311	19,298	24,797
340	4,343	5,533
380	1,286	2,299
420	1,216	2,271

battery storage has a significant effect on the provision of upward and downward flexibility. This is evident by the substantial increment of the upward and downward flexibility compared to the case without the battery storage. Note that the maintenance of high upward and downward operational flexibility is essential in order to compensate the variability of RES.

Table 2. Total upward flexibility (MW).

Minimum flex. (p.u)	0 MWh BES (MW)	60 MWh BES (MW)	120 MWh BES (MW)
<311	3,565	4,536	5,731
340	6,244	8,001	9,186
380	8,869	10,173	11,506
420	10,871	12,187	13,699

Table 3. Total downward flexibility (MW).

Minimum flex. (p.u)	0 MWh BES (MW)	60 MWh BES (MW)	120 MWh BES (MW)
<311	15,254	16,367	17,034
340	14,044	14,899	15,909
380	13,410	14,690	15,876
420	13,220	14,525	15,465

6 Conclusions

In this paper, a unit commitment formulation incorporating a battery storage and a constraint for the system flexibility is proposed. This proposed formulation was used for investigating the impact of different system flexibility levels and the battery storage integration on system flexibility and on the operational cost of a real power system. Results indicate that the provision of a high system flexibility from the conventional units is costly but provides significant amounts of upward flexibility in order to compensate the variability of RES. However, the integration of a battery storage manages to reduce the system operational cost and increase significantly the upward and downward flexibility of the system.

Acknowledgments. This paper is part of the FLEXITRANSTORE project. This project has received funding from the European Union's Horizon 2020 research and innovation programme under grant agreement No 774407.

References

1. Wu, Y., Li, Y., Wu, Y.: Overview of power system flexibility in a high penetration of renewable energy system. In: IEEE International Conference on Applied System Invention (ICASI), Chiba, pp. 1137–1140 (2018)
2. Ma, J., Silva, V., Ochoa, L.F., Kirschen, D.S., Belhomme, R.: Evaluating the profitability of flexibility. In: IEEE PES General Meeting, San Diego, CA, pp. 1–8 (2012)
3. Kiviluoma, J., Rinne, E., Helistö, N.: Comparison of flexibility options to improve the value of variable power generation. Int. J. Sustain. Energy **37**, 761–781 (2018)
4. Helistö, N., Kiviluoma, J., Holttinen, H.: Long-term impact of variable generation and demand side flexibility on thermal power generation. IET Renew. Power Gener. **12**(6), 718–726 (2018)
5. Oree, V., Hassen, S.Z.: A composite metric for assessing flexibility available in conventional generators of power systems. Appl. Energy **177**, 683–691 (2016)
6. Heydarian-Forushani, E., Golshan, M.E.H., Siano, P.: Evaluating the operational flexibility of generation mixture with an innovative techno-economic measure. IEEE Trans. Power Syst. **33**(2), 2205–2218 (2018)
7. Ma, J., Silva, V., Belhomme, R., Kirschen, D.S., Ochoa, L.F.: Evaluating and planning flexibility in sustainable power systems. IEEE Trans. Sustain. Energy **4**(1), 200–209 (2013)
8. Lobato, E., Sigrist, L., Rouco, L.: Use of energy storage systems for peak shaving in the Spanish Canary Islands.In: IEEE PES General Meeting, Vancouver, BC, pp. 1–5 (2013)
9. McPherson, M., Tahseen, S.: Deploying storage assets to facilitate variable renewable energy integration: the impacts of grid flexibility, renewable penetration, and market structure. Energy **145**, 856–870 (2018)
10. O'Dwyer, C., Ryan, L., Flynn, D.: Efficient large-scale energy storage dispatch: challenges in future high renewable systems. IEEE Trans. Power Syst. **32**(5), 3439–3450 (2017)
11. Amoli, N.A., Sakis Meliopoulos, A.P.: Operational flexibility enhancement in power systems with high penetration of wind power using compressed air energy storage. In: Clemson University Power Systems Conference (PSC), Clemson, SC, pp. 1–8 (2015)
12. Yang, J., Zhang, L., Han, X., Wang, M.: Evaluation of operational flexibility for power system with energy storage. In: International Conference on Smart Grid and Clean Energy Technologies (ICSGCE), Chengdu, pp. 187–191 (2016)

Enabling Flexibility Through Wholesale Market Changes – A European Case Study

Ben Bowler[1(✉)], Marcus Asprou[2], Balint Hartmann[3], Peyman Mazidi[4], and Elias Kyriakides[2]

[1] EMAX, Brussels, Belgium
ben.bowler@emaxgroup.eu
[2] University of Cyprus, Nicosia, Cyprus
[3] Budapest University of Technology and Economics, Budapest, Hungary
[4] Loyola Institute of Science and Technology (LoyolaTech), Universidad Loyola Andalucia, Seville, Spain

Abstract. European and global electricity sector decarbonisation is driving changes in wholesale electricity markets as market operators, regulators and system operators encourage increased levels of flexibility. Existing wholesale market design characteristics are changing; new, parallel marketplaces are also emerging. This paper analyses the design space for ancillary services and balancing markets, considers settings in Europe using ENTSO-E annual survey data and discusses changes to market settings that promote flexibility. Finally, it proposes some basic changes to the intraday market.

1 Introduction

The EU 2030 targets impose a 40% reduction of greenhouse gas emission (compared to the 1990 levels). As a result, power systems are experiencing major changes to their infrastructure and operation that increase uncertainty at the point of generation and consumption, affecting the behaviour of all power system participants and their interaction with wholesale electricity markets. In addition, new market designs are required for enhancing and increasing the system flexibility.

With respect to the market design, decarbonisation of the power system creates challenges. For instance, it increases the need to procure services and the need to redispatch or use reserves to balance the electricity system after day ahead market or intraday market gate closure. Decarbonisation also creates opportunities, such as in the intraday market, which provides an opportunity to resolve schedule imbalances that remain at the end of the day ahead market.

Within this paper, options for flexibility in different time frames are discussed, followed by analysis of the design space for ancillary services and balancing markets. The paper then considers the market current market settings in ancillary services markets across Europe.

B. Németh and L. Ekonomou (Eds.): ISH 2019, LNEE 610, pp. 13–22, 2020.
https://doi.org/10.1007/978-3-030-37818-9_2

2 Market Requirements and Options for Flexiblity

The optimal design of the market for increasing system flexibility requires the adjustment of certain parameters of wholesale markets to affect the behaviour and performance of those markets. It is possible to analyse the design space of the market, to consider settings within the design space, and to consider which settings have the greatest impact on flexibility.

Different pathways to a more flexible electricity markets are possible. Efficient market design will find market solutions for the dispatch of flexibility, while reducing the need for TSO-procured reserves. Ultimately, market design should enable price signals for investment in the required flexible resources, and it should ensure that all market participants capable of providing flexibility have access to the market.

Market design must also ensure that the market gives price signals for investment in flexibility. This is best achieved in a market that is free of distortion, such as regulated prices, exit barriers, price caps and undue subsidies. Such distortions endanger long-term market functioning by limiting market participants' ability to obtain sufficient remuneration from the market. While such distortions are not the only factor, they have added to the continuous price decline of European wholesale markets in recent years.

Various steps are proposed in the literature to improve market flexibility. For example, market integration helps to reduce net volatility in supply and demand, decreasing the overall need for flexible resources. By nature, peaks and troughs in two geographically separate electricity systems are to some degree decorrelated. The higher the degree of decorrelation, the lower the combined system volatility and the lower the need for flexible resources. In addition, the total capacity of flexible resources necessary to operate the combined system is lower than for a single system. Reserves can therefore be shared, resulting in potential cost reduction.

Scarcity pricing could enable higher remuneration of flexible resources, if price floors and ceilings were removed. More cost-reflective grid charges can lead to system-beneficial decisions by prosumers. Higher temporal resolution of energy-only market prices can decrease the need for balancing activities of grid operators. At retail level, dynamic pricing and/or aggregation would be a prerequisite for demand participation. Higher geographical resolution of prices as well as grid-adapted price zones could better reflect grid constraints and give a price signal for grid-adaptive investment.

In the next section, the design space for ancillary services, balancing markets, and intraday markets is considered and analysed separately.

3 Flexibility in Ancillary Services Markets

3.1 Market Design Space

There are many local differences between European member states when it comes to implementation of ancillary services. The ENTSO-E annual survey on Ancillary Services Procurement and Electricity Balancing Market Design [1] is used to characterise the design space for implementation of ancillary services. The survey is used to monitor the implementation of the European guideline on electricity balancing and to report on the development towards a European balancing market. The survey has been conducted since 2013; the 2017 survey included results from: Austria, Belgium, Bosnia & Herzegovina, Croatia, Czech Republic, Denmark, Estonia, Finland, France, Germany, Greece, Hungary, Ireland & Northern Ireland, Italy, Latvia, Lithuania, Holland, Norway, Poland, Portugal, Romania, Serbia, Slovakia, Slovenia, Spain, Sweden, Switzerland, and the United Kingdom (Table 1).

Table 1. Requirements for flexibility in different time frames

Market/timeframe	Characteristics	Requirements for flexibility in this timeframe
Balancing/ancillary services	• Covers short term un-expected or unresolved imbalances • Managed by SO • Affected by gate closure time of medium term markets • Includes forward contractual certainty regarding payments – services often include availability fees	• Fast reaction time (typically less than 5 min) • High ramping requirements – speed of variability critical • Low ramping costs • Low dispatch costs • Can be automated based on pre-agreed parameters and contracts • Can operate in short duration – in EU typically up to 1 h • Able to be part of phased response to a system disturbance within clearly defined parameters (typically based around primary, secondary, tertiary response)
Day ahead and intraday markets	• Defined by various products, trading options • Impact on RES/load affected by gate closure times • Increased liquidity reduces reliance on balancing energy requirements • Trading in IDM/DAM results in revenue only associated with the trade – no long term certainty on revenue beyond historical performance • Participation limited by market settings related to bid size and minimum duration	• Predictability – must be able to commit to schedule requirements • Reaction time and ramping requirements governed by gate closure – normally 1 h in EU • Headroom/footroom that allows alteration in committed energy close to real time • Ramping and start up costs are relevant but only if variability forces change in operating characteristics in a timeframe that incurs additional costs • Must be able to respond for minimum duration at a minimum size – sets lower limit on storage or DSR participation, for example

In general, the survey tries to align to the standardised ENTSO definitions for ancillary services, these being: frequency containment reserve (FCR), automatically activated reserve through frequency restoration reserve capacity and energy (aFRR-C/E, previously named regulating power or secondary control reserve), manually activated reserve through frequency restoration reserve capacity and energy (mFRR-C/E, previously named reserve power or tertiary control reserve), and replacement reserve capacity and energy (RR-C/E).

3.2 Analysis of Settings in European Markets

After characterising the design space, the ENTSO-E survey was used to characterise the settings.

First, regarding **resolution**, there have been some downward changes in resolutions for providing capacity in ancillary service markets: a number of countries have moved towards an hourly resolution for mFRR-C, and overall for mFRR-C there seems to be a shift in resolution from higher to lower resolutions. For mFRR-E there is no noticeable shift in resolution – most countries have not changed settings since 2013, with most countries using an hour. Only France and Ireland have set their resolution at 30 min; and only Spain, Belgium and Holland at 15 min.

There has been a general trend towards lower **product resolution** – allowing smaller sources/loads to participate in ancillary services markets. However, there is no clear pattern across borders or between different products. Most are already reasonably small – typically 1–5 MW, although some countries have actually increased their minimum bid size, for example Norway, which has increased bid size from 1–5 MW to 5–10 MW for aFRR-C, and Spain, which has increased from now minimum to >10 MW for aFRR.

Activation time for aFRR-E has changed for a number of countries. France changed from less than 1 min to between 15 min and 1 h. Belgium and Italy changed from between 15 min and 1 h to less than 1 min, and Hungary changed from 5–15 min to less than 1 min.

Regulated pricing for ancillary services is in a minority. Pay as bid is the preferred methodology.

For mFRR-C in 2017 **cost recovery** was mostly from grid users through a tariff. Only Austria and Serbia recover 100% from Balance Responsible Parties, although Norway, Sweden, Finland, and Hungary use a hybrid approach. For mFRR-E only Ireland, Germany and Bosnia Herzegovina use cost recovery from grid users. Almost all other European countries use 100% recovery from BRP (Table 2).

Table 2. Design space variables for ancillary services markets in Europe

Design variable	Variable options
Procurement scheme	Hybrid, Mandatory only, Market only
Product resolution (MW)	Numerical value
Product resolution (time)	Hours, days, weeks, months, years
Distance to real time of auctions	Hours, days, weeks, months, years
Allowed provider type	Generators Only, Generators + Load, Generators + Pump Storage, Generators + Load + Pump Storage, Batteries, Generators + Load + Batteries, Generators + Load + Pump Storage + Batteries
Symmetrical product	Y/N
Pricing mechanism	Pay as bid, Marginal Pricing, Regulated Price
Cost recovery scheme	Grid users, BRP, Hybrid
Monitoring	Real-Time Monitoring, Ex-Post Check, Hybrid
Transfer allowed	Yes, No, Yes only in case of forced outage
Transfer with secondary market	Yes, No
Settlement rules	Pay as bid, Marginal Pricing, Regulated Price, Hybrid
Activation rule	Pro-rata (parallel activation), merit order
Activation time	Time
Partial activation	Yes, in all directions; No in none direction; Only in upward direction; Only in downward direction
Load specific rules	Long term contracts TSO-BSP; Long term auctions; Short term auctions; Specific market solution
TSO control – load participation	No Control; Direct Control (Automatic); Direct Control (Manual); Relay

In ancillary services markets there have been changes made to allow **load and battery participation**, with the greatest change being apparent in FCR. Holland and Switzerland have been the most progressive in allowing **batteries** to participate in ancillary services, but only Switzerland has opened up all markets to batteries. In aFRR and RR the options for batteries are much more limited. Batteries have emerged as allowed participants in FCR-C since 2013 and are now permitted in FCR-C in Finland, Denmark, Holland, Northern Ireland, Ireland, Great Britain, France Switzerland and Germany. The precursor appears to have been the participation of pumped storage in the FCR-C market: all of the countries mentioned already allowed pumped storage in the FCR-C market except Holland and France. Either could be a model for a transition in other countries. For mFRR-C only Great Britain and Switzerland allow batteries, and for RR-C only Switzerland allows batteries.

Load participation is widely accepted for mFRR, but is less common for aFRR. For aFRR-E load was allowed previously in Finland, Denmark and Germany, and is now possible in addition in Belgium, Holland, Austria, and the Czech Republic. Most countries still only allow generation. This may be due to control requirements or settings required for aFRR.

For mFRR-C load was previously allowed in France, Denmark, Slovenia, Hungary, Norway, Great Britain, Germany, Sweden, Finland and Holland. Now load participation is also possible in Belgium, Austria, Slovakia, Ireland, Northern Ireland, and Bosnia and Herzegovina. For mFRR-C, markets allowing access by generators only are in a minority. Load participation is only allowed in a small number of countries for RR. This may be due to the long duration of RR – it can extend for days, making disconnection of load impractical. The requirements for RR are also heavily dominated by generator-only rules.

It should be mentioned that although there are available technologies to provide ancillary services such as inertia and spinning reserve emulation, primary voltage and frequency regulation, there are as yet not many products that could specifically be designed for such services. Where consideration of these services can increase flexibility in power systems, they require updates on the current market platforms.

3.3 Possible Changes: Ancillary Services Markets

First, settings relating to the TSO, and the role of the TSO itself, are important considerations. Service providers should be able to compete with any TSO assets providing system services on an equal footing. A minimum requirement is to increase transparency on the system, including information on cost levels of TSO assets and market participant compensation when they are providing the same services as TSO assets. Article 54 [2] of the proposal for the Electricity Directive states that TSOs shall not be allowed to own, manage or operate energy storage facilities and shall not own directly or indirectly assets that provide system services. It does leave room for derogations for individual countries based on a set of conditions. If derogations are made, they should be on the basis of a level playing field between TSOs and market participants providing the same services. If a power plant can provide the same service more cost-effectively, then consideration should be given to whether it is used instead of the TSO assets. The system needs and cost of providing services by TSO assets should be transparent to reduce information asymmetry between TSOs and market participants. Furthermore, the use of TSO assets should be transparent and provide a price signal and remuneration to market participants providing the same service. In addition, the TSOs should perform an assessment, at regular intervals, of the potential interest of market parties to own the assets in case they can provide the service in a cost-effective manner as also mentioned in the Directive.

Second, marginal pricing for all balancing services, energy and capacity, ensures that service providers receive the full marginal value of the service they are providing. Pay-as-clear pricing is used already in the most relevant market for balancing energy. Pay-as-bid may be justified if there is flexibility in the product definitions formally or in practice, in which case the practice of doing this should be made transparent to all service providers. Marginal pricing should be considered as a more transparent approach for procuring services in ancillary services markets.

4 Flexibility in Balancing Markets

4.1 Market Design Space

There are various approaches defined in the literature that define the design space for the balancing market, and ways to measure flexibility in this timeframe, for example [3–7]. The U.S. Department of Energy Grid Modernisation Laboratory [8] proposes the list of values within Table 3 – **Column A** as metrics for measuring flexibility in balancing markets. Within the EU, Van der Veen (2016) proposed criteria for measuring the effectiveness of balancing markets, broken down according to security of supply, economic efficiency, market facilitation and multinational criteria. The metrics are summarised in Table 3 – **Column B**. The metrics are aligned to objectives stated in the ACER Framework Guidelines (2012), which were adopted as regulations during 2015–2017, and were reinforced by a survey conducted amongst balancing market experts. The ACER guidelines on energy balancing propose the criteria described in Table 3 – **Column C.**

Table 3. Comparison of metrics for measuring effectiveness of balancing market

Colum A	Column B	Column C
U.S. Department of Energy Grid Modernisation Laboratory	Van der Veen (2012)	ACER
1. Loss of load 2. Insufficient ramping 3. Flexibility ratio 4. Wind generation 5. Solar generation fraction 6. Wind generation volatility 7. Solar generation volatility 8. Net load forecasting error 9. Net load factor 10. Maximum ramp rate in net load 11. Maximum ramp capacity 12. Energy storage available 13. Demand response capability 14. Inter-regional transfer capability 15. Intra-regional transfer capability 16. Interruptible tariffs 17. Renewable curtailment 18. Negative LMP 19. Price spikes 20. Load shedding 21. Operational reserve shortage 22. Control performance (CPSs 1. 2; BAAL) 23. Out-of-market operations	1. Security of supply criteria 1.1. Availability of balancing resources 1.2. Balance planning accuracy 1.3. Balance quality 2. Economic efficiency criteria 2.1. Cost allocation efficiency 2.2. Utilisation efficiency 2.3. Price efficiency 2.4. Operational efficiency 3. Market facilitation criteria 3.1. Transparency 3.2. Non-discrimination 4. Multinational criteria 5. Internationalisation costs 6. Social welfare of cross-border exchanges	(a) availability of balancing resources (reserves, availability distribution); (b) total costs of balancing (balancing energy price distribution, activated balancing energy, procurement of reserves); (c) quality of balancing (e.g. Area control error open loop, area control error, unintentional deviations); (d) welfare gain due to cross-border exchanges of balancing energy and reserves

4.2 Possible Changes: Balancing Markets

First, considering steps that are taken to increase price sensitivity to scarcity: the main challenge with sharper scarcity prices is that they increase the price risks for BRPs if there are no tools to hedge their position. In particular, high volatility is more difficult for smaller and non-portfolio market players which can be more exposed to individual price spikes. Moving to single-price imbalance settlement helps to mitigate this issue as market participants can then create instruments to hedge themselves against imbalance price spikes.

Second, considering load curtailment: in the event of involuntary load curtailments, any lost load could be estimated and included in the imbalance positions of individual BRPs and the calculation of the system imbalance volume and the marginal imbalance price.

Third, it may be beneficial to remove the condition that the day-ahead price sets the floor for the up-regulation price and the cap for the down regulation price, and consequently for imbalance prices (except as a backstop price in the event that there are no balancing trades, and if an intraday price cannot be used). Linking the balancing and imbalance prices with the day-ahead prices sets a restriction on free price formation. As updated information is received on e.g. wind forecasts, the situation can change so that the marginal cost to produce energy for up-regulation can be cheaper than it was in the day-ahead stage. Removing this link can lead to situations where the imbalance price is lower than the day-ahead even during up-regulation settlement periods and might provide incentives for market participants to speculate against the imbalance prices. In the long run, free price formation should however lead to the most efficient market outcomes and is in the philosophy of the energy only market.

Fourth, it could be beneficial to remove balancing obligations at the day-ahead and any other stage, better allowing the market to support system balancing. Given the increasing share of fluctuating electricity generation, a rule stipulating that balance must be ensured 12–36 h prior to the hour of operation is contrary to the inherent characteristics of the market. Any balancing obligations at the day-ahead stage, or any other stage, could be removed.

Fifth, better access to information on system imbalance and balancing and imbalance prices in real-time would support market operation. Balancing activations and prices should be published as close to real-time as possible, which serves as an indication for imbalance prices to guide behavior of market participants, allowing them to react in real-time (demand) or submit balancing bids for the next imbalance settlement period (generation). Publishing bid curves could be considered as an additional way to improve transparency, but it requires further analysis.

Sixth, the gate closure to submit bids to the Regulating Power Market should be moved as close to the operating hour as possible. Trading closer to the hour of operation can reduce the forecasting errors by the market participants and contribute to balancing the electricity system as a whole. On the other hand, the closer trading occurs to the delivery hour, the less time there is for TSOs to plan their balancing actions to respond to imbalances and grid congestions.

Lastly, reducing minimum bid size supports new entrants and new behavior, especially from demand-side and other distributed resources such as storage and small-scale generation. The new resources which could emerge would be different in nature to the existing resources. This might require a revision of current requirements to access the balancing market to make sure the requirements are fit for purpose and not overly restrictive towards smaller scale resources. On the other hand, an adequate level of system security needs to be guaranteed and there needs to be a level playing field for different types of resources, e.g. same prequalification process.

5 Flexibility in Intraday Markets

5.1 Proposed Changes: Intraday Markets

In the same manner as in the balancing market some proposed changes can be performed in the intraday market for enhancing system flexibility.

Intraday auctions, first, have proven successful in a number of markets across Europe. Possible settings include market time units for the intraday as close as practically possible to the setting of day-ahead prices and recalculation of available capacities; 15-min products; continuous trading from the resolution of the opening auction until the gate closure time of the ID market; and the possibility of combining with a closing auction in addition to the opening auction. Continuous trading could solve the need of companies with weather-dependent generation or demand, as well as during capacity failures. Extra auctions in addition to an opening auction would be required if there is a relevant market related event that occurs every day. With a large-scale release of reserved capacity, for instance, a further auction could be warranted.

Second, the market would benefit from gate closure time (GCT) as close to real-time as possible. The regional cross-zone intraday GCT should be as close to the operating hour as possible. Shortening the GCT for ID trading should ideally be accompanied with change to the deadline for submitting production plans.

Again, information transparency between market participants and TSOs is important. As market participants trade closer to real time, greater transparency is needed. Market participants should provide the TSO with detailed, up to date information. TSOs should provide up to date, consolidated data on the overall system position. For market participants, this could consist of preliminary and final production plans on a unit level above a certain size limit. This allows the market participants to take better informed decisions on their balance portfolios and the TSOs to prepare for their actions during the balancing timeframe.

Fourth, price cap setting in the intraday market should be set to better reflect the value of lost load.

Fifth, allocation of cross-zone capacity across market timeframes through an explicit cross-zone capacity product could be beneficial to allocate cross-zone capacity across market timeframes, opening up intraday trading between price zones.

Due to increase in distributed generation including renewable resources, reverse power flows in the system can cause technical challenges. These technical challenges can in return affect the prices directly. As a result, possible factors (e.g. penalty) might ensure the limit on the market that is linked to the operation (similar to price cap).

Another point to address this issue could be a consideration of local markets. This could bring higher efficiency in the pricing as not all regions in the system may require the same level of flexibility and reserve. Hence, local flexibility markets could offer local services to overcome the system level challenges. However, remuneration for such services would require further investigations.

6 Conclusions

This paper reviews ancillary, balancing market and intraday settings in Europe. It identifies the bottlenecks and provides a number of proposals for each on how further developments of these markets could provide higher level of flexibility for the power system. The Flexitranstore project, will use simulations and further analysis to assess the viability and impact of these proposals. It will also further develop a definition of the intraday market design space and possible changes to market settings within it.

Acknowledgments. This article was produced as part of the FLEXITRANSTORE project. The FLEXITRANSTORE project has received funding from the European Union's Horizon 2020 research and innovation program under Grant Agreement No. 774407.

References

1. ENTSO-E Market Survey 2017. https://www.entsoe.eu/publications/market-reports/#past-surveys. Accessed September 2018
2. Directive 2003/54/EC of the European Parliament and of the Council of 26 June 2003 concerning common rules for the internal market in electricity and repealing Directive 96/92/EC
3. Riesz, J., Milligan, M.: Designing electricity markets for a high penetration of variable renewables. Wiley Interdiscip. Rev. Energy Environ. **4**, 279–289 (2014)
4. Developing Balancing Systems to Facilitate the Achievement of Renewable Energy Goals, Position Paper, ENTSO-E (2011)
5. Survey on Ancillary Services Procurement and Balancing Market Design, ENTSO-E (2014)
6. Rebours, Y., Kirschen, D., Trotignon, M.: Fundamental design issues in markets for ancillary services
7. Stoft: 2002 Power System Economics - Designing Markets for Electricity
8. https://www.energy.gov/grid-modernization-initiative-0/grid-modernization-lab-consortium

Increasing the Flexibility of Continuous Intraday Markets in Europe

Beáta Polgári(✉), Dániel Divényi, Péter Sőrés, Ádám Sleisz, Bálint Hartmann, and István Vokony

Budapest University of Technology and Economics, 18 József Egry Street, Budapest, Hungary
polgari.beata@vet.bme.hu

Abstract. The FLEXITRANSTORE project inter alia aimed at improving the flexibility of the organized, wholesale electricity markets of Bulgaria and Cyprus fitted into the European electricity market framework. The paper introduces the market settings of these countries and defines the intraday time horizon as a focus for the research. The authors take into account the needs of the different market participants in order to increase the flexibility of the electricity market. After a review and evaluation of the currently available spot market products and orders, the authors propose the introduction of new orders and products on the intraday market so that the trading capabilities of storage and DSM-based technologies are facilitated. The suggestions would help the integration of variable renewable generation as the participation of new flexibility providers such as storage owners and aggregators are facilitated. The most significant proposals of the authors are the volume constrained and the cumulative volume constrained order types that are realized by proposing a new set of execution constraints for existing basket orders. The paper also highlights the advantage of the novel orders compared to the recent product developments of power exchanges.

Keywords: Electricity market · Flexibility · Intraday market

1 Introduction

In line with the EU decarbonisation targets, the share of renewable generation is steadily increasing and the high emission conventional generation is being phased out. The integration of intermittent renewable generation demands more flexibility in the electricity system to balance its scheduling inaccuracy and real-time variability. The assets providing the required flexibility are currently mostly thermal power plants. Unfortunately from this perspective, most of the flexible gas-fired generation is usually priced out of the markets while coal fired power plants are being phased out.

The FLEXITRANSTORE project selected the power markets in Bulgaria and Cyprus to demonstrate and analyse the effectiveness of proposals aiming the smooth integration of South-Eastern Europe to the pan-European electricity market. Apart from the difficulties mentioned above, the authors identified other problems, too, specific to the Bulgarian electricity market. Partly due to the market immaturity, liquidity is

B. Németh and L. Ekonomou (Eds.): ISH 2019, LNEE 610, pp. 23–34, 2020.
https://doi.org/10.1007/978-3-030-37818-9_3

low. The day-ahead market (DAM) started in 2016 while the continuous intraday market (IDM) only in April 2018. The market operator, IBEX, operates a Centralized Market for Bilateral Contracts, besides DAM and IDM [1]. The market platform is provided by Nord Pool. The share of the organized market in Bulgaria is approximately 55%. There is a significant potential for market power as a single company owns around 75% of the built-in generation capacity and more than 80% of the electricity production. Therefore competition is limited. New market participants owning flexible resources would increase competition on the market and support the integration of more renewable energy resources (RES).

Cyprus faces difficulties regarding market competition being an islanded system without any connection to mainland Europe. (The Eurasia interconnector is however planned between Cyprus, Israel and Crete according to [2], but it is only expected to be in operation by December 2023). Moreover, electricity trading is not marketed yet, DAM is not foreseen to be opened before mid-2020 and IDM is expected to start even later. The islanded system also hampers the spread of renewable generation. More flexibility sources available in Cyprus could alleviate these issues.

The paper analyses how new flexibility sources could be incentivized to engage in the electricity markets. First of all, the selection of the time horizon is presented then a connection between market flexibility and liquidity is discussed. Secondly, ideas are gathered for increasing market liquidity. In the following part, the needs of the newly entering market participants demanded by flexibility constraints are explained. The main part of the paper examines the short-term market horizon and the available orders and products to deduce the proposed orders and products, among other things the volume constrained and the cumulative volume constrained orders. Finally, the significance and the further steps of this research are presented after drawing a conclusion.

2 Increasing Market Flexibility

This section contains the deduction of market flexibility improvement starting from the selection of the time horizon, showing the relationship between flexibility and liquidity, ways of improving liquidity and finally satisfying the flexibility needs and capabilities of market participants.

2.1 Focusing on the Intraday Market

The authors decided to put one market time-horizon into the focus of the research. The IDM has been chosen because of the following reasons. Flexibility means fast adjustment capability at short notice. Renewable generation forecasts become more accurate close to delivery period thus possibility for trading should be available to adjust the schedules caused by revised forecasts. A liquid IDM market is suitable for this purpose in contrast to bilateral trading due to time constraints. Proper balancing markets incentivize BRPs to cover their imbalances (and so to avoid high balancing costs), and any open positions before the TSO operated real-time balancing process. The deadline for offer submission can reach 5 min before delivery as in Belgium and in the Netherlands [3]. Day-ahead markets are usually more advanced and liquid than IDMs, and the ability of being flexible is worth more, closer to delivery.

Balancing markets already have several flexibility products, not only standard frequency containment reserves and restoration reserves but also some specific, non-standard ones. For example in the UK specific frequency control ancillary services are introduced that either allow storage and aggregators to participate (in Firm Frequency Response and Enhanced Frequency Response) or are fitted for demand-side management (DSM) based participants (Frequency Control by Demand Management and Demand turn-up, [4]).

Energy storage equipment, especially batteries are in the spotlight of the author's research. The most promising service of batteries clearly is frequency regulation. The current potential for profitable spinning/ramping (RR) reserve service provision is much smaller according to a US study [5]. Batteries are mostly owned by independent power producers or investor-owned utilities. Only high capacity batteries are used for pcak shaving and arbitrage in an outstanding proportion.

An extra argument against relying only on balancing markets for providing flexibility is the longer contract and thus the requirement of constant availability periods (for a whole day or even week) and longer delivery length (usually hours), These specifics are especially unsuitable for storage and DSM assets competing with stored primary fuel based conventional generation. The IDM is close to real time, therefore the bidders do not need to undertake long availability commitments far ahead of the delivery.

The need for flexibility can come from RES generation or from balance responsible parties (BRPs). Our suggestions for more refined market orders are also suitable for the intraday schedule adjustment of traders and RES generators.

Flexibility need will rise to solve grid congestion management issues both on transmission and distribution level. However, this might rather be at a separate new local flexibility market, especially when solving distribution level problems. Some projects are already dealing with the location-related market design such as NODES [6] and ENERY [7]. The scope of the current proposal does not include this spatial dimension, thus neither does any data specify the place of energy injection or off-take in the proposed orders, nor is separate market platform required for the enhanced flexibility trading. Besides, the presented approach of the authors is in line with the current wholesale market designs and market platforms.

2.2 The Connection of Market Flexibility and Liquidity

As mentioned above, market flexibility could be enhanced by attracting new, flexible (storage or DSM-based) participants to the market. Apart from this the market could react to unexpected needs for flexibility if there were corresponding offers every time. These all lead to increasing the number of participants, the offered and traded volume as well as market activity. If a market is liquid, it satisfies more the flexibility needs and price signals are better. Furthermore, a flexible market also gives possibility for conventional flexible power plants to adjust their output in response to the more accurate generation forecast of intermittent renewable generation and load forecast.

Furthermore, this market-centric approach would result in flexibility incentivizing price signals for the considered intraday timeframe.

2.3 Increasing Market Liquidity and Thus Flexibility

Market liquidity and thus flexibility can be increased by market couplings as it attains easier and the most efficient (implicit) access for the available cross-border capacity, increases the number of market participants and orders while diminishes the risk. The XBID (Cross-Border Intraday) project aims this multi-lateral implicit IDM market coupling. The first wave went live in Western-Europe on 12 June, 2018. The Bulgarian market is in the second-wave expected in June 2019 [8].

However, there are transaction types that would result in more trades and thus more social welfare but are currently not supported in coupled markets, only available in single markets. This is the case with cross-product matching used e.g. at the IDM of Hungary (HUPX) but this feature is not supported by the XBID project. This restriction is probably because of algorithmic difficulties and unrepeatability [9]. Cross-product matching allows the matching of different products namely the quarter hourly and hourly products.

Other actions to increase electricity market liquidity can be the introduction of new products (the introduction of block orders is planned in Bulgaria), the reduction of the minimum built-in capacity for compulsory organized market participation (reduced already to 4 MW in Bulgaria) or the procurement of energy previously tendered off the market platform (e.g. system loss). Concerning the latter, Bulgaria is planning to procure the transmission system loss at the power exchange, but the sale of renewable generation under feed-in tariff system done on exchanges could also increase the traded volume. The most prominent marketing place is usually the DAM, but the IDM could also profit from adjusting the day-ahead schedules according to the intraday close-to-delivery weather forecast such as at the HUPX. Opening the futures market in Bulgaria is also under discussion.

Lower limit for minimum offer quantity usually facilitates the participation of new, smaller players but in the case of the IDM trading is based on 0.1 MW lots. The higher resolution of the market orders can also advance liquidity. In most intraday electricity markets, trading is only allowed in hourly products except for the German IDM which can also handle quarter hourly and half hourly products. This enhanced temporal granularity is especially important as balancing is generally settled (and planned to be settled as in [10]) on a quarter hourly basis.

It should also be considered what type of IDM trading is more efficient: the continuous or the auction-based one [11]. Day-ahead markets are typically auction based in Europe while IDMs are mostly continuous. Intraday auctions allowed by market rules are already applied by some countries [12]. In Germany, there is an auction after the main day-ahead auction of 12:00 on D-1 (the day before delivery) at 15:00 exclusively for quarter hourly products. There are separate intraday auctions for the hourly products at D-1 22:00 for the 1–24 h product and at 10:00 for the 12–24 h products [3]. UK also runs a separate auction after the main day-ahead on D-1 at 14:30 solely for half-hourly orders [13]. Some authors state that auctions are more effective than continuous trading. This could be confirmed as most of the trading activity is close to the gate closure. Moreover, at continuous IDMs, an order is matched with the first corresponding order despite that a better matching order would follow later leading to more social welfare.

3 Flexibility Product Development

Based on the above discussions, the authors identify that there is significant opportunity in the development of new flexibility products for the current IDM i.e. products tailored for storage and DSM-based technologies. Before presenting the suggestions, a review is given on some of the European IDMs and special day-ahead products.

3.1 The Review of Current Intraday Markets

The following tables summarize the tradable contracts, order types (Tables 1 and 2), and lead time (Table 3) of some of the European IDMs. The lead time is the gate closure or in other words the end of order submission before delivery ([3, 13–15]).

Table 1. Tradable contracts on intraday markets

Market area	Products			
	Hourly	30 min	15 min	Block
Germany	√	√	√ + D-1 15:00 auction	s, u
Austria	√	–	√	s, u
Belgium & Netherlands	√	–	–	s, u
France	√	√	–	s, u
Switzerland	√	–	√	s, u
Great Britain	√	√ + D-1 14:30, 17:30, D 8:00 auction	–	spec*
Nordic & Baltic	√	–	–	s, u
Hungary	√	–	√	s, u
Bulgaria	√	–	–	s, u

The letters ‘s’ and ‘u’ refer to standardized (base, peak) and user-defined blocks, respectively. Half hourly and quarter hourly (Q) products can help to better balance the portfolios closer to the balancing settlement period. There are also 2 h (2H) and 4 h (4H) long products in Great Britain, as well as special pre-defined blocks by the variation of H, 2H and 4H consecutive products, day base, day peak, day extended peak, day overnight and block 3+4. Extra auctions after the coupled day-ahead auction at 12:00 are also marked in Table 1. These can be interpreted as pre-intraday auctions giving more sequential possibilities for trading.

Table 2. Order types on intraday markets

Market area	Order types		Execution constraints			
	Limit	Market sweep	IoC	FoK	AoN	IBO
Germany	√	√	√	√	√	√
Austria	√	√	√	√	√	√
Belgium & Netherlands	√	–	√	√	–	√
France	√	√	√	√	–	√
Switzerland	√	√	√	√	√	√
Great Britain	√	–	√	√	–	√
Nordic & Baltic	√	–	√	√	–	√
Hungary	√	–	√	√	√	√
Bulgaria	√	–	√	√	√	√

Regarding the abbreviation of order types, IoC means Immediate-or-Cancel, FoK means Fill-or-Kill, AoN stands for All-or-nothing and IBO for iceberg execution constraints. The term 'market sweep order' is used by EPEX for special user-defined blocks with IoC execution constraints [14]. The None (NON) constraint is used in the UK and at HUPX for hourly blocks that allows partial execution, not necessarily executed immediately and can be executed against multiple other orders [15, 16]. As shown in Tables 1 and 2, block orders and the well-known execution constraints are already widely spread including Bulgaria. Nevertheless, current order types and constraints are not adequate for bidding the capability of electric storage or DSM-based bidding. The execution constraints are rather specialized for continuous trading than certain market participants. They limit the validity (IoC and FoK is only valid for immediate execution), divisibility (IoC permits partial acceptance while FoK not) and the visibility of orders (some part of the order are hidden and only one clip is shown).

Lead times closer to delivery could also contribute to the flexible reaction of the market thus both the shorter duration products and the shorter lead times should be encouraged also in the selected countries of this project.

Table 3. Lead time of intraday markets

Market area	Lead time
Germany, Austria, Belgium, Netherlands	5 min
Great Britain, France, Switzerland	30 min
Nordic & Baltic, Bulgaria	60 min
Hungary	90 min

3.2 Flexibility Products on the Day-Ahead Market

Although these are not intraday products, but market platform providers introduced new types of day-ahead orders recently that are worth mentioning. Nord Pool's DSM-related clients can choose the 'flexi order' in the Nordic, Baltic and UK DAMs where Nord Pool is the market operator. The offer maker can give an interval within a day within which the algorithm can choose a shorter specified period for delivery. The algorithm chooses the delivery periods providing the highest welfare [17].

Since 12 December 2018, the loop block has been introduced on the Austrian, Belgian, British, Dutch, French, German and Swiss Day-Ahead markets by EPEX together with the curtailable blocks [18]. The loop block is fitted for storage equipment because a sale and a purchase block can be bonded together as the discharging and charging phase of a battery. The two blocks must be executed or rejected simultaneously. The limitations of this approach are that the offer maker needs to decide the sequence and the possible periods of sale and purchase beforehand limiting the maximum number of charging and discharging periods per day.

Other order types on the day-ahead markets comprise of single hourly and block orders. Blocks can be of linked, curtailable, profile or exclusive types. The two latter are often mentioned as smart blocks on EPEX. Out of these, linked blocks might be used for storage bids and exclusive blocks for DSM bids. For example an off-peak period purchase can be executed on condition of (i) a previous peak sale and (ii) the buy-sale combination being in-the-money but only with precisely defined periods. The exclusive block can find the highest welfare block out of a group to optimize DSM but the block must be predetermined by the bidder [19]. A summary of DAM products is given in Table 4 on the day-ahead tradable contracts per market platform.

Table 4. Tradable contracts on day-ahead markets

Tradable contracts		DAM market platform	
		Nord Pool	EPEX
Single hourly order		√	√
Blocks	Regular	√	√
	Linked	√	√
	Curtailable	√	√
	Profile	√	–
	Exclusive	√	√
	Flexi	√	–
	Big	–	√
	Loop	–	√

3.3 NEW Flexibility Product Suggestions for the Intraday Market

Quarter-Hourly Product and Cross-Product Matching
According to Sect. 3.1, the quarter-hourly product is advised to be introduced on the Bulgarian IDM. It can increase the trades of all participants including RES, storage and conventional power plants. Brijs et al. in [20] also emphasize the importance of closing the positions market-based on the imbalance settlement temporal resolution. It would be beneficial for less flexible participants to be less exposed to balancing prices. Moreover, this could reduce the reserve need of the transmission system and give better price signals for flexibility.

The authors also propose the use of cross-product matching for quarterly limit orders that allows the pairing of an hourly product with corresponding quarter-hourly products.

Exclusive Block Orders
The exclusive block order on the IDM would be similar to the order type known from the DAM, except that not only one but a maximum number of orders could be matched if the limit price criterion is fulfilled. Consequently, the intraday exclusive order would contain a set of bids with volume and limit price parameters plus a third parameter would stand for the maximum number of accepted bids.

Volume Constrained Order for DSM
One of the most refined order types suggested by the authors is a basket order with volume constraint tailored for aggregators bidding DSM. A basket order is a set of orders for different contracts submitted simultaneously to the IDM [21]. Although [22] only proposes the support of linked orders (set of orders with fill-or-kill restriction, all of them shall be executed together) for the common IDM, the basket orders (set of orders without any restrictions) are already implemented in the M7 Trading System widely used in Europe (EPEX, HUPX). On the other hand, according to the requirements for continuous trading matching algorithms in [23], "the algorithm shall be able to support non-standard products to the extent this is technically feasible and approved by the competent regulatory authorities".

The Volume Constrained (VC) order is a conditional order similar to the exclusive block except that not the number of acceptable bids but the total accepted volume is constrained. A VC order contains a set of bids (volume and limit price pairs) for different products plus a parameter that maximizes the daily allocable volume. An example is shown in Fig. 1. Bids can have different volume and price parameters. The cause of non-paired offers marked with light grey can be either the higher offer price (e.g. possibly in QH25) or the volume limit (as in QH32). A modification of this VC order can be when a minimum income constraint (MIC) would be used instead of the limit prices per each bid.

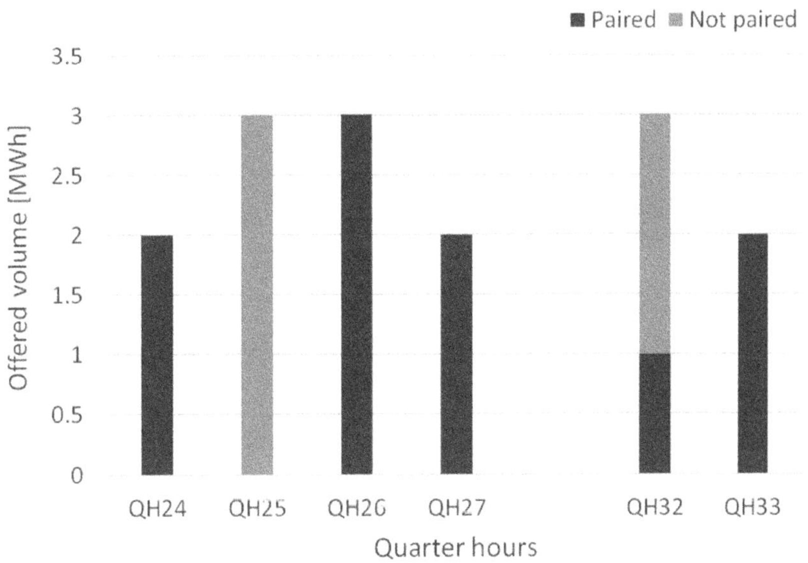

Fig. 1. Example of a Volume Constrained (VC) order.

Given a 10 MWh volume flexible load with a power of 2 MW. The latter determines the volume of the limit purchase bids in the basket. Suppose that 4 MWh energy has been sold previously. Therefore the maximum volume restriction must be 6 MWh. If the Basket Order has been already committed into the order book, the Maximum Volume parameter cannot be modified.

Cumulative Volume Constrained Order for Storage

The other innovative order type suggested by the authors is a basket order with cumulative volume constraint fitted for storage bidding. It considers the state of charge (initial stored energy volume) as a parameter to avoid overcharging or undercharging. Another parameter of the cumulative volume constrained order (CVC) is the maximum cumulative volume that models the maximum power of a battery. The initial volume must be lower than or equal to the maximum volume. Figure 2 presents an example of the CVC order. According to the paired orders, the battery is fully charged in QH24 to 3 MWh, then discharged during QH26-27 limited by the CVC constraint and charged again in QH33 up to 2 MWh. The parameters of the CVC order comply with the recommendations in 20 including charging power, discharging power and limit price.

Suppose a battery with 20 MWh capacity, 0.5 MW charging and 0.4 MW discharging power. The storage capacity gives the maximum cumulative volume restriction while the charging power should be used as the volume of each limit purchase bid in the basket and the discharging power as the volume of the limit sale orders in the basket. The initial volume parameter needs to reflect the charging level of the battery at the beginning of the day. For example a 40% charging level corresponds to 20 MWh · 0.4 = 8 MWh initial volume parameter. If the Basket Order has been already committed into the order book, the Maximum Volume and Initial Volume parameter cannot be modified.

The application of MIC could work also in case of the CVC order and a further development could be the addition of a gradient constraint.

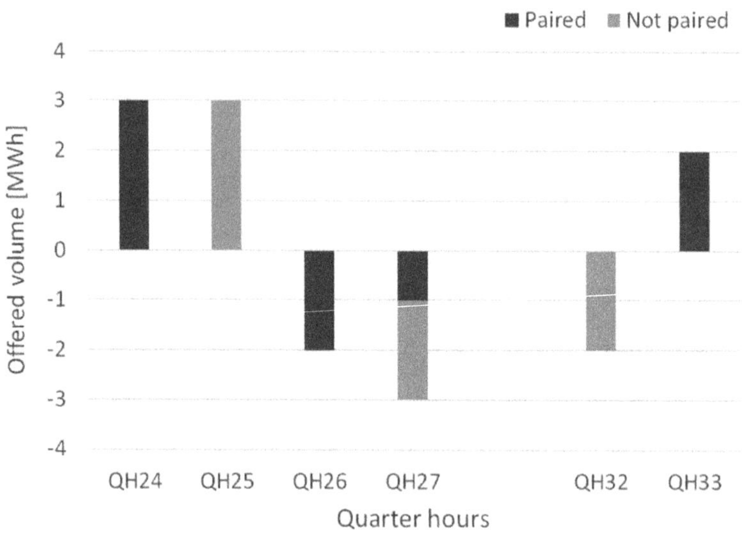

Fig. 2. Example of a Cumulative Volume Constrained (CVC) order.

4 Conclusions

It has been shown that some special products had already been introduced tailored to the needs of new market participants, facilitating their bidding and incentivizing their market participation. But these specific products are set up solely on the day-ahead and balancing markets. The suggestions presented in the paper attempt to fill this market time horizon gap regarding the IDM. The authors recommend establishing quarter-hourly products on the IDM where they are not in use yet (e.g. in Bulgaria) and their extension with cross-product matching. Further recommendations include the introduction of the volume constrained orders and cumulative volume constrained orders tailored for aggregated DSM and storage bidding, respectively. These new order types would also be useful at any other European continuous IDMs. The added value is significant as the authors did not find any other flexibility product proposals in the literature for IDMs.

The next phase of the research is the functional extension of the current continuous IDM algorithm by the above mentioned orders. The initial prototyping results are promising. The final goal of the project is the demonstration of the products on a parallel-run marketplace in Bulgaria and Cyprus.

Acknowledgments. This work has been developed at Budapest University of Technology and Economics within the boundaries of FLEXITRANSTORE project, which is an international project. FLEXITRANSTORE (An Integrated Platform for Increased FLEXIbility in smart TRANSmission grids with STORage Entities and large penetration of Renewable Energy Sources, grant agreement ID: 774407) aims to contribute to the evolution towards a pan-European transmission network with high flexibility and high interconnection levels.

References

1. IBEX – URL. http://www.ibex.bg/en. Accessed 24 Feb 2019
2. Eurasia Interconnector. https://www.euroasia-interconnector.com. Accessed 24 Feb 2019
3. Nord Pool AS: Product Specifications, Belgian and the Netherlands, German, Nordic and Baltic, French, Austrian market areas, 12 June 2018
4. Balancing Services - National Grid. https://www.nationalgrideso.com/balancing. Accessed 24 Feb 2019
5. U.S Energy Information Administration: U.S. Battery Storage Market Trends, May 2018. https://www.eia.gov/analysis. Accessed 24 Feb 2019
6. NODES unveils innovative market design for energy flexibility-trading, 7 November 2018. https://www.nordpoolgroup.com/message. Accessed 25 Feb 2019
7. First trade on flexibility platform enera completed successfully, 5 February 2019. http://www.epexspot.com/en/press
8. Juhász, G.: What is XBID and what are the expected updates. In: HUPX Workshop, September 2018
9. Association of European Energy Exchanges (EUROPEX): Proposal for a common set of requirements used for the continuous trading matching algorithm (2017). Point 1.v
10. Imbalance Settlement Period (Electricity Balancing Market). https://www.emissions-euets.com/-imbalance. Accessed 1 Mar 2019
11. Weber, C.: Adequate intraday market design to enable the integration of wind energy into the European power systems. Energy Policy **38**, 3155–3163 (2010)
12. Commission Regulation (EU) 2015/1222 of 24 July 2015 establishing a guideline on capacity allocation and congestion management, Article 63. https://eur-lex.europa.eu. Accessed 24 Feb 2019
13. Nord Pool AS: Product Spec., GB market area, 29 January 2019
14. EPEX spot product info. http://www.epexspot.com/en/product-info. Accessed 25 Feb 2019
15. Market rules of HUPX, v11.0, valid from 1 March 2019
16. All NEMOs' proposal for products that can be taken into account by NEMOs in intraday coupling process in accordance with Article 53 of the Commission Regulation (EU) 2015/1222 of 24 July 2015 establishing a guideline on capacity allocation and congestion management, 13 November 2017
17. Nord Pool: flexi order description. https://www.nordpoolgroup.com/Flexi-order. Accessed: 26 Feb 2019
18. EPEX spot introduces curtailable blocks and loop blocks on all day-ahead markets. https://www.epexspot.com/en/press. Accessed: 26 Feb 2019
19. EPEX SPOT: Introduction of Smart block bids – Linked block orders and Exclusive block orders, January 2014
20. Brijs, T., Jonghe, C., Hobbs, B.F., Belmans, R.: Interactions between the design of short-term electricity markets in the CWE region and power system flexibility. Appl. Energy **195**, 36–51 (2017)
21. NEMO Committee: Continuous Trading Matching Algorithm Public Description. https://www.ote-cr.cz/xbid/public-description. Accessed: 26 Feb 2019
22. Association of European Energy Exchanges (EUROPEX): Intraday Products Proposal (2017)
23. EUROPEX: Proposal for a common set of requirements used for the continuous trading matching algorithm (2017)

An Improved Flexibility Metric Based on Kernel Density Estimators Applied on the Greek Power System

Konstantinos F. Krommydas[1,2(✉)], Akylas C. Stratigakos[1,2], Christos Dikaiakos[2], George P. Papaioannou[2], Elias Zafiropoulos[3], and Lambros Ekonomou[3]

[1] Department of Electrical and Computer Engineering, University of Patras, 26504 Rion, Patras, Greece
krommydas@ece.upatras.gr
[2] Research, Technology and Development Department, Independent Power Transmission Operator (IPTO) S.A., 89 Dyrrachiou & Kifisou Str. Gr, 10443 Athens, Greece
[3] Institute of Communications and Computer Systems, 9 Iroon Polytechniou Street, 15780 Athens, Greece

Abstract. The large-scale integration of variable renewable energy (VRE) in power systems, such as wind and solar, increases the flexibility needed to maintain the load-generation balance. In order for the power operators to plan for secure and reliable operation, they must examine whether there exists sufficient power system flexibility to meet ramps caused by the increased VRE integration and the system demand. In this context, the paper aims to propose an improved flexibility metric to accurately evaluate the flexibility level of a power system in the planning stage. The proposed metric is based on kernel density estimators and expresses the probability of the flexibility residual (the difference between the available flexibility and the expected net load ramps) being less than zero. The Greek power system is used as case study in order to evaluate the proposed index. In particular, the unit commitment optimization problem with flexibility constraints for ten different scenarios based on the ENTSO-E methodology for the time period 2020–2024 is solved and then the proposed metric is calculated. Finally, this index is compared to the well known insufficient ramping resource expectation (IRRE) metric to further evaluate it.

Keywords: Flexibility metrics · Unit commitment problem · Kernel function

1 Introduction

The increasing penetration of variable renewable energy (VRE) presents significant technical challenges for balancing the power system supply and demand. More specifically, as the integration of VRE increases, the net load, which is the difference between electrical load and renewable power becomes more variable with steeper ramps, shorter peaks, and lower turn-downs than the load itself [1]. Therefore, system flexibility or the ability of a system to meet changes in demand and VRE production plays a crucial role [2].

B. Németh and L. Ekonomou (Eds.): ISH 2019, LNEE 610, pp. 35–46, 2020.
https://doi.org/10.1007/978-3-030-37818-9_4

Until now, traditional capacity adequacy studies focused on the amount capacity available, the forced outage rate of each resource and the system demand. In these studies, metrics such as loss of expectation (LOLE) [3, 4], the expected energy not served (EENS) or well-being analysis [5] are standard measures by which a planned portfolio is evaluated and have served the industry well until now. However, as the shares of VRE increases, the planning approach will need to change and should incorporate new metrics and additional factors, such as the availability and ramp rate of the resources, start-up time and minimum stable operation of the conventional generating units, the variability and uncertainty of the net load and reserve provision planning. But assessing flexibility is not yet common practice. Although indirectly involved in the planning of most power systems, such assessments do not yet follow a comprehensive, transferable methodology that might facilitate common practice among operators. It is a highly complex process considering many operation and design aspects [6].

In this frame, different flexibility studies and metrics have been presented in the literature that focus on the long-term planning of power systems. In [7] a methodology is developed to deal with a significant degree of uncertainty about time and location of generating assets expansion and in [8] a method is proposed for evaluating the system flexibility to adapt quickly with limited costs to every change from the initial planning conditions, with a particular regard to the changes in generation. In [9] a flexibility metric is constructed that reflects the largest variation range uncertainty that the system can sustain by using the four-element framework (time, uncertainty, action and cost), while in [10] the insufficient ramping resource expectation (IRRE) metric is proposed to measure power system flexibility for use in long-term planning and is derived from traditional generation adequacy metrics. In [11] the metric of the number of periods of flexibility deficit (PFD) is proposed to characterize the flexibility of a system.

Taking into account the new flexibility metrics and approaches presented in the literature, the conventional adequacy studies of the Greek power system are revisited by taking into account flexibility requirements, resources and metrics. In particular, considering ten different scenarios for the Greek net load for the time period 2020–2024 that are determined based on the ENTSO-E methodology, we solve the unit commitment optimization problem by including both system power balance constraints and ramp capability constraints of the flexible resources. Furthermore, the flexibility metric IRRE is calculated, which expresses the number of up and down ramps for which there would be insufficient ramping resources available. Calculation of the IRRE follows a similar structure to the LOLE, however, it targets on the unavailability of flexibility and not on the shortage of generation capacity.

However, the IRRE metric does not take into account the temporal correlation between the flexibility available and the flexibility required which may lead to misleading results about the system's flexibility. For example at the highest net load levels, most generator would be at high output and available to offer the additional downward flexibility likely to be required by the system, therefore IRRE underestimates the system flexibility which can lead to incorrect planning and unnecessary investments from the system operator.

In order to overcome this drawback, we utilize the PFD index approach. In particular, we calculate the flexibility residual, which is the difference between the available flexibility and the expected net load ramps, for each observation and horizon and then determine the probability of the flexibility residual distribution being less than zero. However, in contrast to the PFD approach, we use a non-parametric kernel distribution to estimate the probability distribution function, which can provide more accurate and robust results. Furthermore, the proposed metric incorporates the temporal correlation between the flexibility available and the flexibility required in comparison to the IRRE index. Overall, the main contribution of this work can be summarized as follows:

1. The traditional adequacy studies of the Greek power system are upgraded by adding the flexibility requirements and flexibility resources.
2. Using a Kernel density estimator we estimate the flexibility of the system over different time horizons and directions and highlight the time horizons of net load ramping in which the system is most vulnerable.
3. Simulations that use real data for both the flexibility requirements and flexibility resources are performed for solving the unit commitment optimization problem with flexibility requirements constraints of the Greek power system for the time period 2020–2024.

The rest of the paper is organized as follows. In Sect. 2, some preliminaries in flexibility of power systems are introduced. In Sect. 3, the flexibility metrics that are used in the flexibility assessment study of Greece are presented. In Sect. 4, the unit commitment problem is formulated as a mixed integer linear program. In Sect. 5 simulation results are shown and, finally, in Sect. 6 some conclusions are drawn.

2 Preliminaries in Flexibility

Before we describe the flexibility metrics, firstly, we present some preliminaries in power system flexibility as defined in [10]. Also, in Table 1 the nomenclature is displayed.

2.1 Net Load Ramps

The net load ramp time series $NLR_{t,i}$ for time interval i, at observation t, is defined as follows

$$NLR_{t,i} = NL_t - NL_{t-i}. \tag{1}$$

The net load ramps can be positive or negative. Also, the two following variables are introduced

$$NLR_{t,i,+} = NLR_{t,i} \forall NLR_{t,i} > 0 \tag{2}$$

$$NLR_{t,i,-} = -NLR_{t,i} \forall NLR_{t,i} < 0 \tag{3}$$

which denote positive and negative ramping occurrence.

Table 1. Nomenclature.

Symbol	Definition
G	The set of the 26 generators
S	Scenario
T	The length of generation planning horizon (24 h)
g	Indices of thermal generators
t	Time period
SU_{gt}	Start up cost of unit g at time t
SD_{gt}	Shut down cost of unit g at time t
L_g	Minimum ON time of generator g
l_g	Minimum OFF time of generator g
$SUtime_g$	Start up time of generator g
$SDtime_g$	Shut down time of generator g
NL_t	The net load at observation t
P_g^{max}	Maximum power generation capacity for generator g
P_g^{min}	Minimum power generation capacity for generator g
RU_g	Ramping up limit of generator g
RD_g	Ramping down limit of generator g
RS_t	Spinning reserve requirement at time t
S_g^{max}	Maximum spinning reserve of generator g
D_t	Forecasted hourly demand at time t
$VOLL$	Value loss of load (Euro/MWh)
a_g, b_g	Linear coefficients of fuel cost function
$a_g^{'}, b_g^{'}$	Linear coefficients of reserve cost function
u_{gt}	Commitment decision for unit g at time t
v_{gt}	Binary variable, start up action of unit g at time t
w_{gt}	Binary variable, shut down action of unit g at time t
p_{gt}	The thermal power generation output of unit g at time t
s_{gt}	Spinning reserve of unit g at time t
Δ_t	Load-shedding loss at time t

2.2 Available Resource Flexibility

Given the operating state of a unit g we can calculate the available upward and downward flexibility it can offer. In particular, the available upward (+) flexibility ($Flex_{t,g,i,+}$), for a resource g, over a horizon i, at observation t, is defined as

$$Flex_{t,g,i,+} = RU_g * (i - (1 - u_{gt}) * SUtime_g). \tag{4}$$

Upward flexibility is bounded by the maximum production capacity of the resource. Also, when the resource is offline, its start up time must be less than the considered time horizon, and also it must have sufficient time to reach minimum stable production. These two aforementioned conditions can be expressed mathematically as follows

$$p_{gt} + Flex_{t,g,i,+} \leq P_g^{\max}. \tag{5}$$

$$p_{gt} + Flex_{t,g,i,+} \in \mathbb{R} - (0, P_g^{\min}). \tag{6}$$

The downward (−) flexibility available ($Flex_{t,g,i,-}$) from a unit g can be in a similar manner calculated as

$$Flex_{t,g,i,-} = RD_g * u_{gt}. \tag{7}$$

The downward flexibility available is constrained by the minimum stable production level, i.e. the unit must either operate at least at minimum production level or shut down completely. The constraints for the downward flexibility available can be expressed mathematically as follows

$$p_{gt} - Flex_{t,g,i,-} \geq 0 \tag{8}$$

$$p_{gt} - Flex_{t,g,i,-} \in \mathbb{R} - (0, P_g^{\min}). \tag{9}$$

2.3 System Flexibility

Addition of the available flexibility of each individual unit results in the total system flexibility

$$Flex_{t,SYSTEM,i,+/-} = \sum_{\forall g} Flex_{t,g,i,+/-}. \tag{10}$$

3 Flexibility Metrics

Now, we are ready to introduce the flexibility metrics utilized in this work.

3.1 IRRE

The available flexibility distribution ($AFD_{i,+/-}(X)$) can be calculated with the empirical cumulative distribution function of the flexibility available X. The $AFD_{i,+/-}(X)$ represents the probability that X MW, or less, of flexible resource will be available during the i time horizon. The probability that the system has insufficient ramp resources at each observation, over each time horizon and direction, are calculated from the cumulative probability. In order to exclude these cases when the flexibility available is equal to the net load changes, the net load ramps are reduced to a value just below the net load value (for example 1 MW). Therefore, the insufficient ramping resource probability (IRRP) at observation t, over a horizon i, is:

$$IRRP_{t,i,+/-} = AFD_{i,+/-}(NLR_{t,i,+/-} - 1). \tag{11}$$

The sum of the IRRP values over the entire time series, $T_{+/-}$, for each direction results in the IRRE, as defined in the following equation

$$IRRE_{i,+/-} = \sum_{\forall t \in T_{+/-}} IRRP_{t,i,+/-}. \tag{12}$$

3.2 The Proposed Metric

Given the total available flexibility of the power system and the flexibility requirements calculated from the net load ramps, we can calculate the flexibility residual of the system at each observation t and time horizon i, for both directions (upward and downward) as follows

$$FlexResidual_{t,SYSTEM,i,+} = Flex_{t,SYSTEM,i,+} - \max\{NLR_{t,i,+}, 0\} \tag{13}$$

$$FlexResidual_{t,SYSTEM,i,-} = Flex_{t,SYSTEM,i,-} - \max\{NLR_{t,i,-}, 0\}. \tag{14}$$

In order to calculate the probability distribution function of the flexibility residual time series we fit a non-parametric kernel distribution [12]. This way we avoid making any assumption about the underlying distribution of the time series. Given a random sample $x_1, x_2, \ldots, x_n$ from an unknown distribution, the kernel density estimator is given by

$$\widehat{f_h(x)} = \frac{1}{nh}\sum_{i=1}^{n} K\left(\frac{x - x_i}{h}\right) \tag{15}$$

where $\widehat{f_h(x)}$ the estimated probability density function, K the kernel function, n the sample size and h the bandwidth. The kernel function and the bandwidth are the parameters that need to be selected. For this study we selected the Gaussian Kernel, which is the most often used and is given as follows

$$K(x) = \frac{1}{\sqrt{2\pi}}\exp\left(-\frac{1}{2}x^2\right). \tag{16}$$

The parameter h needs tuning in order to avoid over fitting or under fitting the underlying time series, which was selected according to Silverman's rule of thumb [13]. Finally, by calculating the probability of the flexibility residual being less than zero, we determine the probability that the system has insufficient ramp resources.

4 Unit Commitment Problem Formulation

In this section we provide a brief description of the unit commitment problem and the constraints included in this study. The problem is formulated as a mixed integer linear program (MILP), programmed using the YALMIP toolbox [14] and solved with a commercial solver. To minimize the total generation cost of the system the following deterministic unit commitment problem for energy and ancillary services co-optimization is formulated as in [15] where the objective is the following

$$\begin{aligned} \min \sum_{g \in G} \sum_{t \in T} (SU_{gt} v_{gt} + SD_{gt} w_{gt}) + \sum_{t \in T} \sum_{g \in T} [(b_g p_{gt} + a_g u_{gt}) + (b'_g s_{gt} + a'_g u_{gt})] \\ + VOLL \sum_{t \in T} \Delta_t \end{aligned} \tag{17}$$

which minimizes start-up costs, shut-down costs and generation costs, using a linear approximation of the fuel cost function. Equations (18) and (19) represent the constraints that affect the unit commitment decision. In particular, the minimum ON and OFF time constraints are given by

$$\begin{aligned} & u_{gt} - u_{g(t-1)} \leq u_{g\tau} \, \forall g \in G, \, t \in T, \, \tau = t, \ldots, \min\{t + L_g - 1, |T|\} \\ & u_{g(t-1)} \quad u_{gt} \leq 1 \quad u_{g\tau} \, \forall g \in G, \, t \in T, \, \tau - t, \ldots, \min\{t + l_g - 1, |T|\} \end{aligned} \tag{18}$$

where τ is a possible operating time period starting from time t, while the start up and shut down action constraints can be expressed as

$$\begin{aligned} & v_{gt} \geq u_{gt} - u_{g(t-1)} \, \forall g \in G, \, t \in T \\ & w_{gt} \geq - u_{gt} + u_{g(t-1)} \, \forall g \in G, \, t \in T. \end{aligned} \tag{19}$$

The level of generation and spinning reserve provided by each unit is determined by the following constraints $\forall g \in G, \, t \in T$

$$\sum_{g \in G} p_{gt} + \Delta_t \geq D_t \tag{20}$$

$$P_g^{\min} u_{gt} \leq p_{gt} \leq P_g^{\max} u_{gt} \tag{21}$$

$$p_{gt} \geq 0 \tag{22}$$

$$p_{gt} - p_{g(t-1)} \leq P_g^{\min}(2 - u_{gt} - u_{g(t-1)}) + RU_g(1 + u_{g(t-1)} - u_{gt}) \tag{23}$$

$$p_{g(t-1)} - p_{gt} \leq P_g^{\min}(2 - u_{gt} - u_{g(t-1)}) + RD_g(1 - u_{g(t-1)} + u_{gt}) \tag{24}$$

$$\sum_{g \in G} s_{gt} \geq RS_t \tag{25}$$

$$0 \leq s_{gt} \leq S_g^{\max} u_{gt}. \tag{26}$$

Equations (20) and (25) are constraints concerning the total system generation and reserve based on the net load forecast, while the remaining constraints are based on techno-economic parameters of each plant, including generation output, ramp rate and reserve provision. For this study we assumed that the net load is perfectly forecasted in each scenario and that perfect market completion exists, thus the generation bids represents the marginal cost of each unit. Also, only spinning reserves were considered.

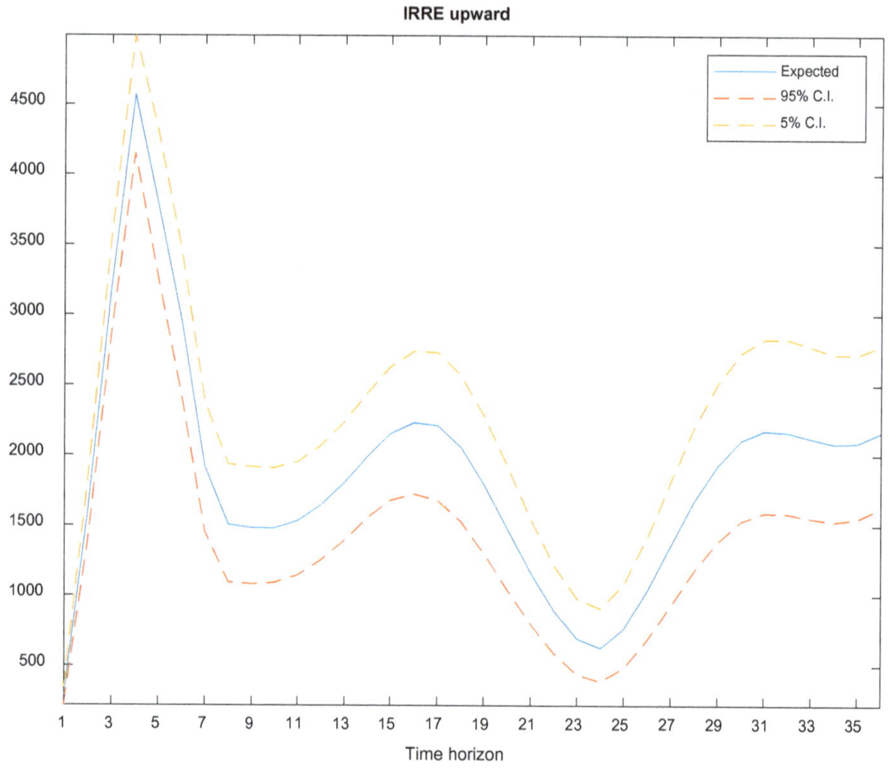

Fig. 1. The expected value of the IRRE metric and its confidence interval for the time period 2020–2024 for the Greek power system.

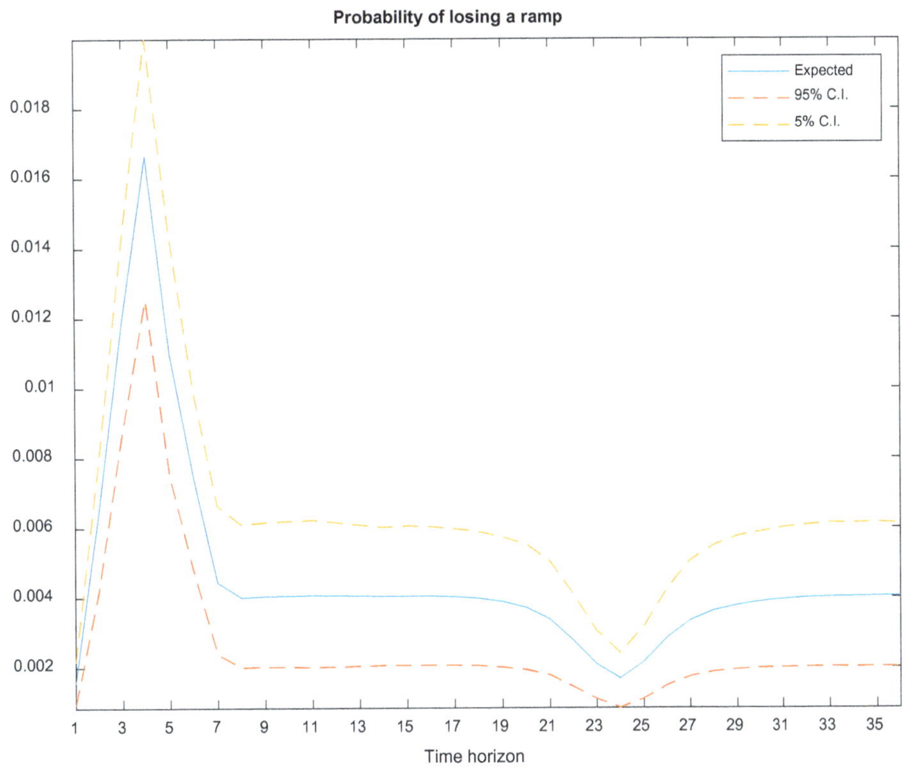

Fig. 2. The expected value of the proposed metric and its confidence interval for the time period 2020–2024 for the Greek power system.

5 Results

For the case study we used data from the Greek power system which are utilized in the Mid-term Adequacy Forecast conducted by ENTSO-E [16]. The different demand forecast scenarios examined in the case study correspond to various climate years for the time period 2020–2024, while the base case scenario both in demand growth and VRE penetration and normal scenario for hydro production are assumed. These scenarios are determined based on the methodology provided in [16].

The operation of thermal generators is modeled in detail, using all the techno-economic data available (ramp rates, generation constraints, marginal costs, etc.), while the maintenance schedule of thermal units considered for this study is the same as the schedule assumed in the adequacy studies of the Greek power system. Furthermore, commission of new power plants and expected retirement of old plants are also taken into account based on the Greek ten year network development plan. The operation of the hydro units is implemented for peak shaving operation for each scenario under consideration. The contribution of interconnections is regarded as the equivalent of a thermal unit of 500 MW base load with 95% availability. Although this is a conservative approach, since Greece is highly dependent from interconnections, this will enable us to remain on the safe side. Finally, the demand side management, since it is considered a safety measure and does not participate in the market, it is used only in cases of emergency as a last resort in the simulation study.

After taking into account all the examined scenarios, the expected values of the IRRE index and the proposed metric for different time intervals are determined along with two extreme values of the confidence interval. The results for both metrics for the case of upward ramps for the time period 2020–2024 are shown in Figs. 1 and 2. It is evident that the calculated metrics produce similar results, as they both identify the 5-h time horizon as the most vulnerable to lose a ramping event. By comparing the rest of the time horizons, the IRRE displays also a larger value in the 16-h time horizon in contrast to our proposed metric. This can be attributed to the fact that IRRE does not take into consideration the temporal correlation between the ramping events and the available flexibility at the particular observation, but rather compares them ex-post, which in turn may lead to falsely identifying particular time horizons as more in risk than they actually are. For example, in the time frame from 5:00 am to 9:00 pm the power system will experience the full extent of variability in the net load, resulting in very large upward ramps, but also in that time frame the system is able to provide a lot of upward flexibility since most units at 5:00 am are operating at the technical minimum (base units) or are offline (peak units). This underlines the benefit of modeling the probability of losing a ramp based on the flexibility residual.

Furthermore, as one can observe from Fig. 1, the expected value of the IRRE metric (i.e. 4500) is used only to highlight the time horizons that are at most risk rather than to determine the absolute level of risk. On the contrary, the metric proposed in this paper provides realistic values of the probability of the system having insufficient ramp resources in the different time horizons. In particular, our methodology shows that there is a probability ranging from under 0.012 to over 0.020 of losing an upward ramp in the 5-h horizon for the various scenarios.

6 Conclusions

In this paper, an index based on Kernel density estimators is proposed that can measure accurately the flexibility of a power system. Calculating the metric and comparing it with the IRRE index in the unit commitment problem for the Greek power system, we showcased that the proposed metric achieves to evaluate effectively the flexibility of the system in the long term horizon. Therefore it can be a very useful tool for system operators in the planning phase and can accommodate flexibility studies.

Acknowledgments. The paper includes partial results on flexibility assessment for transmission systems in line with the FLEXITRANSTORE project. The authors are grateful to all the other partners cooperating in the project and the support received from them. This project has received funding from the European Union's Horizon 2020 research and innovation programme under grant agreement No. 774407, for which the authors wish to express their gratitude. For further information, please check the website: www.flexitranstore.eu.

References

1. Ma, J., Silva, V., Belhomme, R., Kirschen, D.S., Ochoa, L.F.: Evaluating and planning flexibility in sustainable power systems. IEEE Trans. Sustain. Energy **4**(1), 200–209 (2013)
2. International Energy Agency: Harnessing Variable Renewables, pp. 41–67. International Energy Agency, Paris (2011)
3. Calabrese, G.: Generating reserve capacity determined by the probability method. Trans. Amer. Inst. Elect. Eng. **66**(1), 1439–1448 (1947)
4. Billinton, R., Allan, R.: Reliability Evaluation of Power Systems, 2nd edn. Plenum, New York (1996)
5. Billinton, R., Fotuhi-Firuzabad, M.: A basic framework for generating system operating health analysis. IEEE Trans. Power Syst. **9**(3), 1610–1617 (1994)
6. Stratigakos, A.C., Krommydas, K.F., Papageorgiou, P.C., Dikaiakos, C., Papaioannou, G.P.: A suitable flexibility assessment approach for the pre-screening phase of power system planning applied on the greek power system. In: Proceedings of IEEE EUROCON 2019 - 18th International Conference on Smart Technologies, Novi Sad, Serbia, pp. 1–6 (2019)
7. Bresesti, P., Capasso, A., Falvo, M.C., Lauria, S.: Power system planning under uncertainty conditions. Criteria for transmission network flexibility evaluation. In: Proceedings of 2003 IEEE Bologna Power Tech Conference, Bologna, Italy, July 2003
8. Capasso,A., Falvo, M.C., Lamedica, R., Lauria, S., Scalcino, S.: A new methodology for power systems flexibility evaluation. In: Proceedings of 2005 IEEE Russia Power Tech, St. Petersburg, pp. 1–6, June 2005
9. Zhao, J., Zheng, T., Litvinov, E.: A unified framework for defining and measuring flexibility in power system. IEEE Trans. on Power Syst. **31**(1), 339–347 (2016)
10. Lannoye, E., Flynn, D., O'Malley, M.: Evaluation of power system flexibility. IEEE Trans. Power Syst. **27**(2), 922–931 (2012)
11. Lannoye, E., Flynn, D., O'Malley, M.: Assessment of power system flexibility: a high-level approach. In: Proceedings of 2012 IEEE Power and Energy Society General Meeting, San Diego, CA, pp. 1–8 (2012)
12. Bowman, A.W., Azzalini, A.: Applied Smoothing Techniques for Data Analysis. Oxford University Press Inc., New York (1997)
13. Silverman, B.W.: Density Estimation for Statistics and Data Analysis. Chapman & Hall/CRC, London (1986)
14. Lofberg, J.: YALMIP: a toolbox for modeling and optimization in MATLAB. In: Proceedings of IEEE International Symposium on Computer Aided Control Systems Design (2004)
15. Huang, Y., Pardalos, P.M., Zheng, Q.P.: Electrical power unit commitment: deterministic and two-stage stochastic program-ming models and algorithms. Springer (2017)
16. ENTSO-E: Mid-Term Adequacy Forecast 2018 Edition, Brussels (2018)

Zero Renewable Incentive Analysis for Flexibility Study of a Grid

Peyman Mazidi[1(✉)], Gregory N. Baltas[1], Mojtaba Eliassi[1], Pedro Rodriguez[1,2], Ricardo Pastor[3], Michalis Michael[4], Rogiros Tapakis[4], Vasiliki Vita[5], Elias Zafiropoulos[5], Christos Dikeakos[6], and George Boultadakis[7]

[1] Loyola Institute of Science and Technology (LoyolaTech), Loyola University Andalucia, Seville, Spain
pmazidi@uloyola.es
[2] Research Center on Renewable Electrical Energy Systems (SEER), Technical University of Catalonia Barcelona, Barcelona, Spain
[3] Centro de Investigação em Energia, REN - State Grid, S.A, Sacavém, Portugal
[4] Transmission System Operator, Cyprus (TSOC), Strovolos, Cyprus
[5] Institute of Communications and Computer Systems, 9 Iroon Polytechniou Street, 15780 Athens, Greece
[6] Independent Power Transmission Operator (IPTO TSO), Athens, Greece
[7] European Dynamics Luxembourg SA, 12, Rue Jean Engling, 1466 Luxembourg, Luxembourg

Abstract. Power systems with renewable sources are undergoing various changes to safely accommodate higher share of renewables. In this process, they require different changes ranging from technical characteristics to regulations. In order to manage technicality of the integration of the renewables, storage units play an important role. On the other hand, many countries allocate incentives to the green sources of energy. In this paper, we develop a flexibility-oriented day-ahead market model that accounts for renewable sources and storage units where no incentive is provided to the renewable sources. Additionally, a flexibility indicator (FLEXIN) is adopted to demonstrate the provided flexibility to the system. FLEXIN is defined by accounting for five sources of flexibility, reserve from conventional power sources, available renewables, storage units, transmission line availability and active demand. The results demonstrate how a system can cope with renewable sources with no incentive in the presence of storage. This paper is part of the Horizon 2020 Flexitranstore project.

Keywords: Battery energy storage system · Flexibility · Power system

Nomenclature

Indexes

i, j	Index of network buses
h	Time interval (hour)
g	Index of generating units

B. Németh and L. Ekonomou (Eds.): ISH 2019, LNEE 610, pp. 47–60, 2020.
https://doi.org/10.1007/978-3-030-37818-9_5

Sets

Ω_{ℓ}^{i}	Set of all buses connected to bus i
Ω_{TH}^{i}	Set of all thermal generating units connected to bus i
Ω_{WIND}^{i}	Set of all wind generating units connected to bus i
Ω_{BESS}^{i}	Set of all BESS units connected to bus i

Parameters

α_g	Incremental production cost (\$/MWh^2)
b_g	Linear production cost (\$/MWh)
c_g	Fixed operation cost of generator g (\$/h)
r_g	Reserve cost of generator g (\$/h)
VOLL	Value of lost load (\$/MWh)
VOEMS	Value of emissions (\$/Kg)
EMS_g	Emissions for generator g (Kg/MWh)
$D_{h,i}^{al}$	Proposed active power demand in bus i at time h (MW)
ζ_{min}/ζ_{max}	Minimum and maximum active load variability (%)
P_g^{max}	Maximum limit of active power generation of unit g (MW)
P_g^{min}	Minimum limit of active power generation of unit g (MW)
$\bar{R}_g$	Ramp-up limit of generator g (MW/h)
$\underline{R}_g$	Ramp-down limit of generator g (MW/h)
η_c	Efficiency of charging (%)
η_d	Efficiency of discharging (%)
$P_{i,j}^{max}$	Maximum active power flow of branch connecting bus i to j
$P_{i,j}^{min}$	Minimum active power flow of branch connecting bus i to j
γ	Spinning reserve (%)
ρ_g	Startup cost of generator g [\$]
σ_g	Shutdown cost of generator g [\$]
$X_{i,j}$	Reactance between lines i and j [Ω]
δ^{max}	Maximum voltage angle
δ^{min}	Minimum voltage angle

Variables

$u_{h,g}$	Commitment status of generator g (1: on, 0:off)
$y_{h,g}$	Startup decision of generator g at time h (1:up, 0:no change)
$z_{h,g}$	Shutdown decision of generator g at time h (1:down, 0:no change)
$P_{h,g}$	Active power generated by unit g at time h (MW)
$D_{h,i}^{drP}$	Realized demand in bus i at time h (MW)
$R_{h,g}$	Reserve at time h by generator g (MW)
$P_{h,i,j}$	Active power flow of branch connecting bus i to j at time h (MW)
$D_{h,i}^{lsP}$	Load shedding in bus i at time h (MW)
$P_{h,g}^{c}$	Active power charge into BESS g at time h (MW)
$P_{h,g}^{d}$	Active power discharge from BESS g at time h (MW)
$\delta_{h,i}$	Voltage angle of bus i at time h (rad)
$SOC_{h,g}$	State of charge of BESS g at time h

1 Introduction

The current power system is still based on the past technologies where large-scale fossil fuel based power plants produce majority of the electricity share in the generation mix and consumers do not have significant participation [1]. Thus, the electricity market of the next decade will move towards a decentralized structure with more variable generation and improved technologies and increased consumer participation. On the other hand, in the balancing markets with centralized and decentralized (self-scheduling) structures, if there is a security threat to the system due to locational imbalances, the centralized market can solve such an issue where additional factors such as reserve and congestion are also simultaneously considered. Hence, there are still some European countries that consider a centralized system, such as Greece [2].

As renewables' production continues to increase in Europe, power systems and electricity markets face multiple challenges in order to successfully integrate large amounts of non-dispatchable renewable energy sources (RES). Figure 1 shows a perspective of the electricity generation outlook in the European Union (EU) per technology in a scenario that incorporates existing policies and announced policy intentions [3].

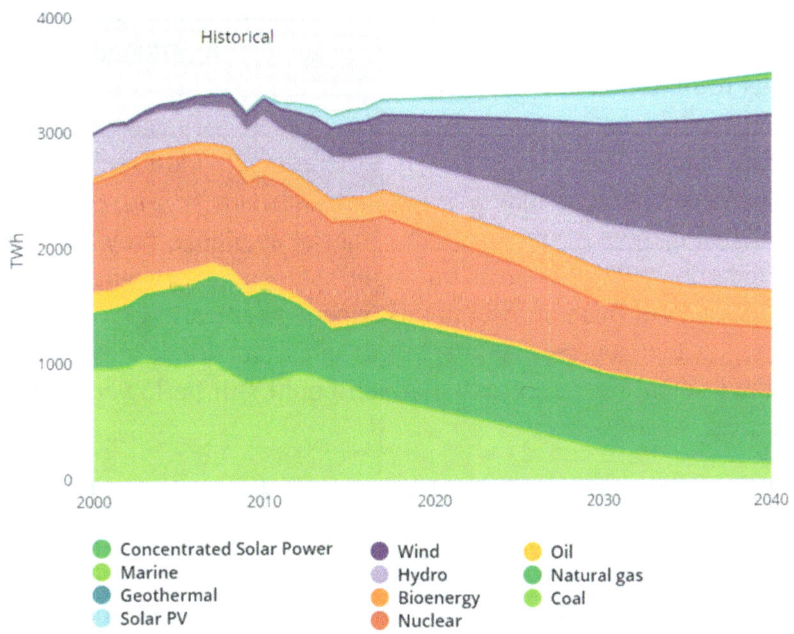

Fig. 1. Electricity generation outlook in the EU per technology. Source (IEA/WEO 2018) [3].

The political goals on the supporting mechanisms clearly state that renewable power plants must be integrated in the electricity market to create an open, competitive, transparent and non-discriminatory environment among all players [4]. Therefore, the economic sustainability of renewable generation in the electricity market is critical for pursuing the new energy policy in the EU. Without further incentives and with the end of the contracts for subsidized renewable production (e.g. feed-in-tariffs), renewable producers will be forced to participate under the same market rules as any other producer [5].

One key issue associated with RES is the stochastic nature of sources. High variability of RES requires an analogous degree of flexibility from the available resources to maintain power balance. At the same time, renewable producers are being challenged about how they can contribute to the system's flexibility. Advances in renewable generation technologies (e.g. smart maintenance [6]) are allowing them to provide more flexibility to the system. Thus, the provision of flexibility services is a key, both for renewable producers, as they can sell additional services and increase their revenue, and for the system operators, as they increase the list of flexibility sources that are available for the system's operation.

2 Assumptions and Structure

It is very common to consider that competitive markets without market failures can be modeled through an optimization programming where market surplus is maximized [7]. The model proposed in this paper for flexibility studies covers a day-ahead market operation as a social welfare maximization concept. In this regard, system operator makes the decisions in order to have the lowest operation cost in the next 24 h horizon. This is achieved by a mixed integer nonlinear programming problem (MINLP) where the objective is to minimize the operation cost of the system.

Renewables have incentives in many countries where they can receive subsidies for generation. This reduces the costs and assists them to provide low bids to the market. One other form of incentive is that renewable sources are prioritized over other types and this means that if there is any renewable energy available, they must be produced, regardless of the impact that this decision might have on the system, e.g. increase costs or create congestion. For this purpose, in this paper, specifically we do not consider any incentive for renewable sources. As a result, it is assumed that they participate in the market with their actual operating costs, which could still be lower than other participating generators.

2.1 Reserve

There are various methods to model reserve in a power system. Different operators consider different approaches based on their system requirements and perspectives [8]. In this paper, the secondary control reserve is defined as capability of committed generators to provide additional production in case it is required. On the other hand, the minimum reserve is considered as a percentage of system load at any time. Although it is assumed that the generator must have enough capacity to provide additional support, this excess support is not limited by the ramp characteristics of the generator. As a result, if a generator is already running and has free capacity, the free capacity is accounted for as the reserve. This simplification of neglecting the ramping limitation and solely considering committed capacity might result in solutions that are cheaper, easier to solve (due to mathematical difficulties) with unrealistic reserve values.

2.2 Flexibility Index

Uncertainty, outage and variations can be considered as three main origins that raise the need for flexibility in a system. Uncertainty in the system can be defined as coming from various sources such as intermittent renewables. In a similar manner, outages of generators can significantly impact the system. Finally, variations in the demand could change the condition of the system. These parameters could show their influence in the ranges from real-time to medium term and long term. In other words, the system requires flexibility in control, operation and planning. In this paper, we try to address the flexibility from operation point of view, Fig. 2.

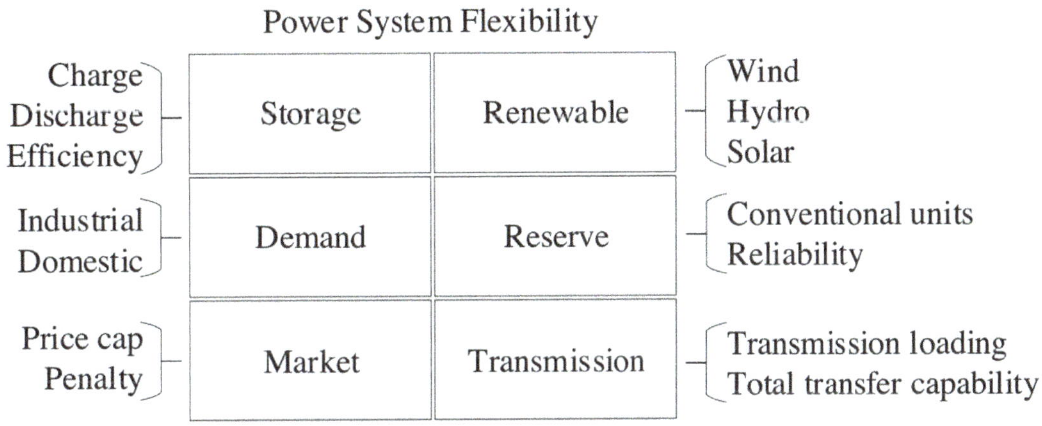

Fig. 2. Flexibility factors in a power system [9]

The three hierarchical levels that the flexibility can come into play are generation, transmission and distribution. Renewable sources and battery energy storage system (BESS) can produce at the generation level to provide flexibility. For the distribution, active load can provide flexibility by shifting the demand from one time to another. In the same way, a BESS could also play such a role in form of an active distribution node (ADN). On the transmission level, loading of the transmission lines could be linked to the flexibility of the lines. For instance, if the lines in an area of the system are mostly fully loaded, the flexibility of the system from transmission point of view is very low. Since each transmission system operator is responsible for monitoring its grid, they calculate such concept as available transfer capability (ATC) [10]. Additionally, the ATC can be shared with the generation companies so that they can consider such flexibility in their operation planning as well. Considering these points, we also adopt the flexibility definition as a flexibility index (FLEXIN) provided by [9] to show the degree of system flexibility.

3 Modelling

The proposed model is formulated through an optimization problem that is explained below. In the following, the case studies are defined and the input data of the test system are mentioned.

3.1 Mathematical Formulation

The mathematical formulations for the model have an MINLP form where the objective is to minimize operation cost of the whole system.

$$\begin{aligned}
&\sum_{h,g \in \Omega^i_{TH}, \Omega^i_{WIND}} \left(\alpha_g P^2_{h,g} + b_g P_{h,g}\right) \\
&+ \sum_{h,g \in \Omega^i_{BESS}} \left(\alpha_g {P^c_{h,g}}^2 + b_g P^c_{h,g} + \alpha_g {P^d_{h,g}}^2 + b_g P^d_{h,g} + c_g SOC_{h,g}\right) \\
&+ \sum_{h,g \in \Omega^i_{TH}} \left(r_g R_{h,g}\right) + \sum_{h,g \in \Omega^i_{TH}, \Omega^i_{WIND}} \left(c_g u_{h,g}\right) + \sum_{h,g \in \Omega^i_{TH}, \Omega^i_{WIND}} \left(\rho_g y_{h,g}\right) \\
&+ \sum_{h,g \in \Omega^i_{TH}, \Omega^i_{WIND}} \left(\sigma_g z_{h,g}\right) + \sum_{h,i} \left(VOLL \times D^{lsP}_{h,i}\right) \\
&+ \sum_{h,g \in \Omega^i_{TH}} VOEMS \times EMS_g \times P_{h,g}
\end{aligned} \tag{1}$$

The operation cost of the system is presented in (1). It includes the production cost of conventional generators, wind generators and BESS in the first two terms. Cost of reserve is accounted for as well as commitment, startup and shutdown costs. Final two terms include the penalty for load shedding and cost of emissions. Subsequently, the system constraints are formulated as follows:

$$\begin{aligned}
&\sum_{g \in \Omega^i_{TH}, \Omega^i_{WIND}} P_{h,g} + D^{lsP}_{h,i} - D^{drP}_{h,i} - \sum_{g \in \Omega^i_{BESS}} P^c_{h,g} + \sum_{g \in \Omega^i_{BESS}} P^d_{h,g} \\
&= \sum_{j \in \Omega^i_{\ell}} P_{h,i,j} \qquad \forall h, i
\end{aligned} \tag{2}$$

$$u_{h,g} P^{min}_g \le P_{h,g} \le u_{t,g} P^{max}_g \qquad \forall h, g | g \in \Omega^i_{TH,WIND} \tag{3}$$

$$P_{h,g} - P_{h-1,g} - \bar{R}_g \le 0 \qquad \forall h, g | g \in \Omega^i_{TH}, \Omega^i_{WIND} \tag{4}$$

$$P_{h-1,g} - P_{h,g} - \underline{R}_g \le 0 \qquad \forall h, g | g \in \Omega^i_{TH}, \Omega^i_{WIND} \tag{5}$$

$$u_{h+1,g} - u_{h,g} = y_{h+1,g} - z_{h+1,g} \qquad \forall h \langle N_h, g | g \in \Omega^i_{TH}, \Omega^i_{WIND} \tag{6}$$

$$SOC_{h,g} = SOC_{h-1,g} + P^c_{h,g} \eta_c - \frac{P^d_{h,g}}{\eta_d} \qquad \forall h, g | g \in \Omega^i_{BESS} \tag{7}$$

$$P_g^{min} \leq P_{h,g}^c \leq P_g^{max} \qquad \forall h, g | g \in \Omega_{BESS}^i \tag{8}$$

$$P_g^{min} \leq P_{h,g}^d \leq P_g^{max} \qquad \forall h, g | g \in \Omega_{BESS}^i \tag{9}$$

$$P_g^{min} \leq SOC_{h,g} \leq P_g^{max} \qquad \forall h, g | g \in \Omega_{BESS}^i \tag{10}$$

$$SOC_{h,g} = 0 \qquad \forall h = 1, N_h, g | g \in \Omega_{BESS}^i \tag{11}$$

$$P_{h+1,g}^c - P_g^{max} + SOC_{h,g} \leq 0 \qquad \forall h \langle N_h, g | g \in \Omega_{BESS}^i \tag{12}$$

$$P_{h+1,g}^d - SOC_{h,g} \leq 0 \qquad \forall h \langle N_h, g | g \in \Omega_{BESS}^i \tag{13}$$

$$P_{h,i,j} = \frac{\delta_{h,i} - \delta_{h,j}}{X_{i,j}} \qquad \forall h, i, j | j \in \Omega_\ell^i \tag{14}$$

$$-P_{i,j}^{max} \leq P_{h,i,j} \leq P_{i,j}^{max} \qquad h, i, j | j \in \Omega_\ell^i \tag{15}$$

$$\delta^{min} \leq \delta_{h,i} \leq \delta^{max} \qquad \forall h, i \tag{16}$$

$$\delta_{h,i} = 0 \qquad \forall h, i = slack\ bus \tag{17}$$

$$\sum_{g \in \Omega_{TH}^i} R_{h,g} \leq \sum_{g \in \Omega_{TH}^i} \left(u_{h,g} P_g^{max} - P_{h,g} \right) \qquad \forall h \tag{18}$$

$$\sum_{g \in \Omega_{TH}^i} R_{h,g} \geq \gamma \sum_i D_{h,i}^{drP} \qquad \forall h \tag{19}$$

$$\sum_{h,i} D_{h,i}^{drP} = \sum_{h,i} D_{h,i}^{al} \tag{20}$$

$$(1 - \zeta_{min}) \sum_i D_{h,i}^{al} \leq \sum_{h,i} D_{h,i}^{drP} \qquad \forall h \tag{21}$$

$$\sum_{h,i} D_{h,i}^{drP} \leq (1 + \zeta_{max}) \sum_i D_{h,i}^{al} \qquad \forall h \tag{22}$$

$$0 \leq D_{h,i}^{lsP} \leq D_{h,i}^{drP} \qquad \forall h, i \tag{23}$$

Equation (2) shows the demand balance where transmission lines are considered. As a result, we obtain locational marginal prices in each node of the system as Lagrange multiplier of this equation.

Equation (3) limits minimum and maximum production level of conventional and wind generators. (4) and (5) set the upward and downward ramping capabilities of the generators. (6) provides the link between startup, shutdown and commitment status of the generators.

Equations (7)–(13) are related to BESS. (7) shows the relation between charge, discharge and state-of-the-charge of the BESS by accounting for charge and discharge efficiencies. While (8)–(10) limit maximum and minimum charge, discharge and SOC values, (11) considers that the value of the SOC of the BESS should be the same at the beginning and the end of the simulation period. (12) considers that the amount that the BESS can be charged at any time should be less than the difference between the maximum capacity of the BESS and the SOC that is already available at the previous hour. Similarly, (13) considers that the amount of discharge of the BESS at any time cannot be more than the available amount of SOC at the previous time.

Equation (14) defines the power flow between the buses to be linearly related to the difference between the voltage angles of the buses, where (15) and (16) sets the limit on the maximum and minimum capacity of the lines and the voltage angles of the buses. (17) assumes that the voltage angle of the slack bus is zero.

Equation (18) considers that participation of each conventional generator in the hourly defined reserve in this system should be less than the maximum available capacity of that generator. The reserve should also be greater than some percentage of the realized load, (19).

Equations (20)–(23) consider the demand side management where moved sum of the demand in the simulation horizon should be equal to the sum of the forecasted demand. The demand flexibility degree is also considered as a percentage where the moved demand can only be a percentage of the forecasted demand, (21), (22). In the end, maximum load shedding at any time and bus should not be more than the realized load (23).

The mathematical formulation for FLEXIN is also provided in (24), where conventional generators provide flexibility through their ramping and reserve, BESS provides flexibility along with renewable sources individually, active demand from the customer side and ACT from transmission side also deliver flexibility.

$$\begin{aligned}&\sum_{g\in\Omega^{i}_{TH}} R_{h,g} + \sum_{g\in\Omega^{i}_{WIND}} \left(P_{g}^{max} - P_{h,g}\right) + \sum_{g\in\Omega^{i}_{BESS}} \left(P_{g}^{max} - SOC_{h,g}\right)\\&+\sum_{i,j}\frac{\left(P_{i,j}^{max} - \left|P_{h,i,j}\right|\right)}{2} + \zeta_{max}\sum_{i} D_{h,i}^{al} \qquad \forall h\end{aligned} \tag{24}$$

3.2 Case Studies

In order to investigate the performance of the model and analyze flexibility of the the system, different case studies have been defined and carried out, Table 1. The main idea to design the case studies was to link the FLEXIN as a metric for evaluating flexibility of the system and demonstrating influence of various sources of flexibility. While Case1 considers a basic case without the main flexibility sources, Case5 puts together all the defined flexibility sources.

Table 1. Considered study cases for the analyses

Cases	Reserve	Renewable	Storage	Active demand
Case1	×	×	×	×
Case2	√	×	×	×
Case3	√	√	×	×
Case4	√	√	√	×
Case5	√	√	√	√

The model is coded in GAMS software v25.0.3 [11] and solved with BARON solver v17.10.16 [12] on a computer with 64 GB of RAM and i7-7820X @3.60 GHz Intel processor. As for the test system, we considered modified IEEE 24-bus test system [13].

3.3 Input Data

Penalty for performing a load shedding is considered by defining the value of lost load (VOLL) parameter as 10000 ($/MWh). The value of emissions (VOEMS) for conventional generators is considered to have a cost of 20 ($/kg). Charge and discharge efficiencies of BESS are assumed to be 98%. 0.9 and 1.1 in radian are the minimum and maximum allowable values for voltage angles at each bus. The reserve has a value of 20% with respect to the demand (γ), which means the system reserve at any time should be more than or equal to the 20% of the system demand. Finally, demand flexibility at each bus assumed to be 4% (ζ). The rest of the data about the test system can be found in [9]. The model considers three BESS units, two wind plants and twelve non-renewable sources.

It should be reminded that the wind generators are connected to the grid in buses 5 and 9 and the three BESS units are connected to the grid in buses 10, 13 and 18. The value of demand at hour 20:00 in bus 5, 9, 10, 13 and 18 is 70, 124, 146, 212 and 242 MW, respectively. The connecting transmission lines for the renewable generator buses are: L3 (b1–b5), L6 (b3–b9), L8 (b4–b9), L9 (b5–b10), L12 (b8–b9), L14 (b9–b11), L15 (b9–b12). Similarly, the connecting transmission lines for the BESS buses are: L9 (b5–b10), L10 (b6–b10), L13 (b8–b10), L16 (b0–b11), L17 (b10–b12), L18 (b11–b13), L20 (b12–b13), L22 (b13–b23), L29 (b17–b18) and L31 (b18–b21).

4 Results

A comparison between the maximum and minimum changes of prices and loading of the lines are provided in Table 2. As it can be seen, addition of reserve in Case2 has resulted in similar prices as in Case1, where no reserve is accounted for. Two explanations can be provided here. The first explains that in Case1, the system is already providing enough reserve, which by explicitly considering the 20% reserve, the operation cost of the system does not change significantly.

Table 2. Price and line loading summary

Cases	Price ($/MWh)		Transmission loading (MW)	
	Max	Min	Max	Min
Case1	31.24	28.18	1048.74	508.99
Case2	31.20	27.64	1253.20	535.29
Case3	29.66	27.20	1026.62	563.25
Case4	30.88	27.25	942.30	456.63
Case5	30.62	27.22	1209.39	650.41

The other explanation is that this type of formulating reserve, (18) and (19), does not accurately display the impact of reserve on the operation cost of the system. This is also in line with the previous explanation as this kind of formulation neglects the characteristics of the generators from ramping perspectives. Therefore, it can be concluded that the reserve formulation in [9] provides a more realistic model for reserve. However, as mentioned earlier, the current formulation has simpler form as it avoids the Big-M formulation. The impact of this formulation on the rest of the case studies should be taken into account when analysing the results.

Another observation from Table 2 is the distinct impact of renewables on the prices in Case3. One interesting reflection is the reduction in loading of the lines when storage enters the operation in Case4. Both maximum and minimum loading values have been reduced, where similar prices as in Case3 are obtained. This shows how adding storage can relieve some stress from transmission lines in overall. Finally, by bringing in demand flexibility, although the prices see a reduction, the loading of the lines increases. It should be mentioned that there is no cost for loading of the lines in the operation cost.

Figure 3 illustrates different loading levels of the transmission lines at peak hour, 20:00 with the value of 2233 MW. It should be mentioned that in Case1, none of the flexibility sources are considered and Case5 considers that only four of the flexibility sources have direct impact on the operation cost. The only flexibility factor that is not considered in the operation cost is the loading of the lines. Thus, it is not expected that the loading of the lines are minimized.

Comparing the two cases in Fig. 3 and considering the values in Table 2, it can be seen that adding the sources of flexibility to the system causes a decrease in the operation cost and an increase in the loading of the transmission system. Furthermore, it can be seen that except for L8 and L12, the loading of the mentioned lines increases by considering new flexibility sources.

Moreover, it can also be seen that except for L9, L13 and L22, the loading of the lines have reduced. As a result, in short, out of sixteen related lines to the flexibility sources, nine lines have had decrease in their loading values after consideration of flexibility sources. It should however be mentioned that the overall loading of the transmission system has increased. Thus, the flexibility sources have provided some loading relief (congestion management) for the connected buses at the peak hour.

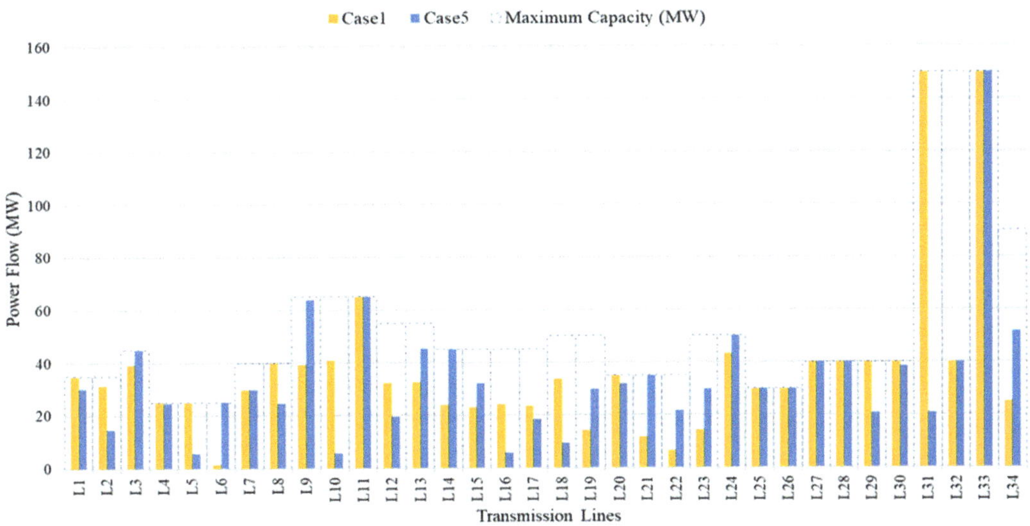

Fig. 3. Transmission line loading for Case1 and Case5.

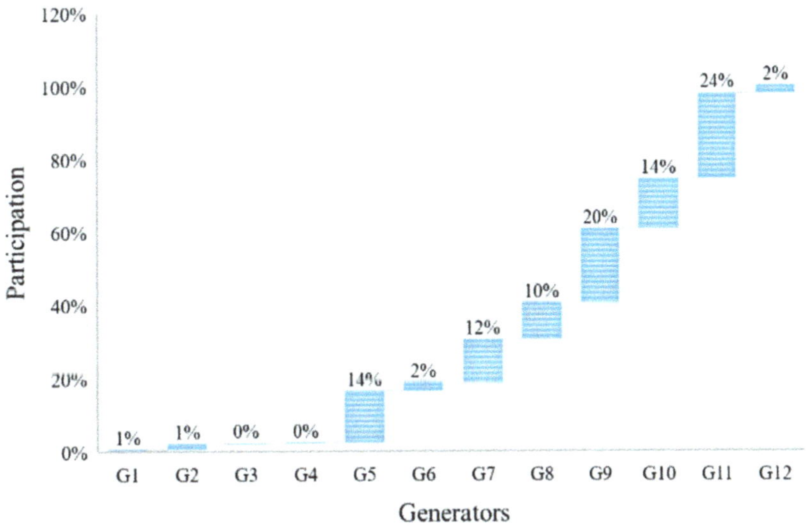

Fig. 4. Production share in Case2.

Figure 4 illustrates the generation percentage from generators, with G11 having the highest value. G11 is a conventional generator with maximum capacity of 440 MW. G11, G9 and G5 are the largest generators having about 60% of the production share. The 0% for G3 and G4 is due to rounding of the values.

Figure 5 displays the operation of batch of BESS in the system where all sources of flexibility are considered. It can be seen that the BESS starts charging when there is large amount of renewables and discharge in high load hours (h9, h10) and near peak and peak hours (h13, h14, h20, h21) (arbitrage condition). It is also interesting to see that the BESS is out of operation solely for four hours in the whole day.

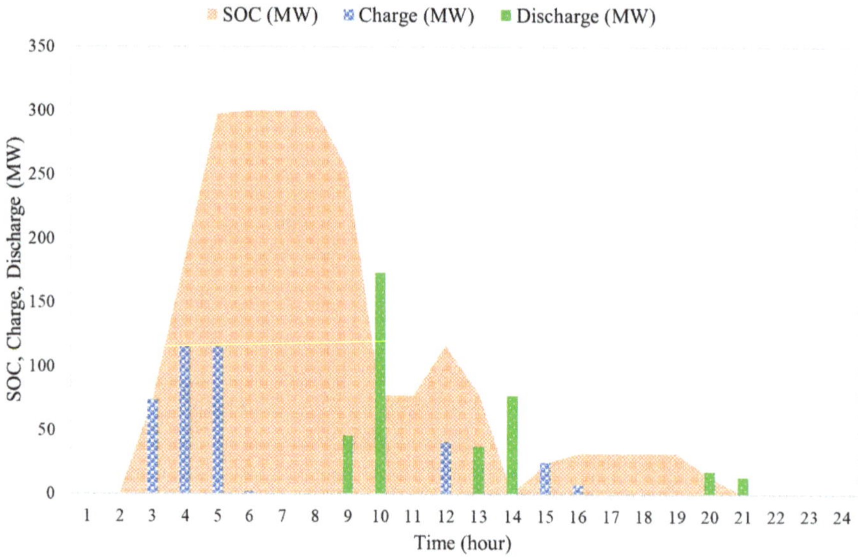

Fig. 5. Operation of aggregated BESS in Case5.

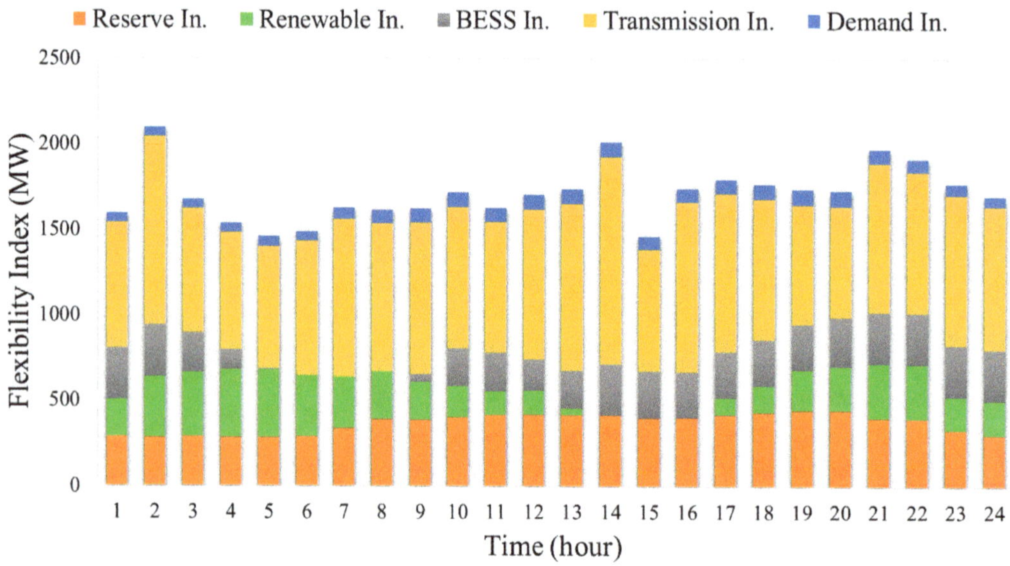

Fig. 6. FLEXIN for Case5.

Finally, Fig. 6 illustrates the hourly system flexibility provided from the considered five flexibility sources. It can be seen that on this particular day, the maximum and minimum system flexibility are calculated to be 2096 MW (at hour2) and 1456 MW (at hour5). The lowest value of flexibility occurs at hour5 resulting from three consecutive charging periods by the BESS. Now, this can be considered as a reference study for the operator. For instance, the operator can later look back, check the operating conditions on this day, and see whether they were satisfactory. If not, the operator can modify the system in order to achieve better system flexibility and security. Considering that there are various parameters affecting the flexibility, the operator could particularly ask BESS to be in standby at hour5 in case there is a contingency.

5 Conclusions

This paper incorporates five sources of flexibility into a day-ahead operation in power systems where the incentive for renewable sources are ignored. The paper also considers a simplified formulation for the reserve and the results show the impact of such simplification on the operation cost and flexibility. Finally, the adopted flexibility index displays how a system operator could investigate flexibility of the grid over a future operating horizon and make required adjustments to improve the operation and hedge against contingencies. Future studies could discuss the sensitivity of change in the operation cost and flexibility of the system with respect to various values of incentive for renewable sources. Additionally, self-scheduling aspect could also be adapted in order to demonstrate the possible modifications in case the decentralized approach is selected.

Acknowledgments. This work was supported by the European Commission under project FLEXITRANSTORE—H2020-LCE-2016-2017-SGS-774407 [14] and by the Spanish Ministry of Science under project ENE2017-88889-C2-1-R. Any opinions, findings and conclusions or recommendations expressed in this material are those of the authors and do not necessarily reflect those of the host institutions or funders.

References

1. Proposal for a regulation of the European parliament and the council on the internal market for electricity. European Commission, Brussels (2017)
2. Supporting document for the network code on electricity balancing: European network of transmission system operators for electricity (ENTSOE), Brussels (2014)
3. IEA/World Energy Council: World Energy Outlook 2018 (2018). https://www.iea.org/weo/
4. European Commission: COMMUNICATION FROM THE COMMISSION - Clean Energy for All Europeans, 30 November 2016. https://eur-lex.europa.eu/legal-content/EN/TXT/?qid=1481278671064&uri=CELEX:52016DC0860
5. Imperial College London, Nera Economic Consulting, DNV GL: Integration of renewable energy in Europe. Final report for the European Commission (2014)
6. Mazidi, P.: From condition monitoring to maintenance management in electric power system generation with focus on wind turbines, KTH Royal Institute of Technology, Comillas Pontifical University, Delft University of Technology, p. 222 (2018)
7. Kazempour, J., Hobbs, B.F.: Value of flexible resources, virtual bidding, and self-scheduling in two-settlement electricity markets with wind generation—part I: principles and competitive model. IEEE Trans. Pow. Syst. **33**(1), 749–759 (2018)
8. Ela, E., Milligan, M., Kirby, B.: Operating reserves and variable generation. National Renewable Energy Laboratory (NREL) (2011)
9. Mazidi, P., Baltas, G.N., Eliassi, M., Rodriguez, P.: A model for flexibility analysis of RESS with electric energy storage and reserve. In: 7th International Conference on Renewable Energy Research and Applications (ICRERA), Paris (2018)

10. Christie, R.D., Wollenberg, B.F., Wangensteen, I.: Transmission management in the deregulated environment. Proc. IEEE **88**(2), 170–195 (2000)
11. GAMS Development Corporation: General Algebraic Modeling System (GAMS), Washington, DC, USA (2018)
12. Tawarmalani, M., Sahinidis, N.V.: A polyhedral branch-and-cut approach to global optimization. Math. Program. **103**(2), 225–249 (2005)
13. IEEE Commmitte Report: IEEE reliability test system. IEEE Trans. Power Apparatus Syst. **PAS-98**(6), 2047–2054 (1979)
14. FLEXITRANSTORE: An Integrated Platform for Increased FLEXIbility in smart TRANSmission grids with STORage Entities and large penetration of Renewable Energy Sources, European Union's Horizon 2020 research and innovation programme 2017–2021

Conflict of Interests Between SPC-Based BESS and UFLS Scheme Frequency Responses

Mojtaba Eliassi[1(✉)], Roozbeh Torkzadeh[1], Peyman Mazidi[1], Pedro Rodriguez[1,2], Ricardo Pastor[3], Vasiliki Vita[4], Elias Zafiropoulos[4], Christos Dikeakos[5], Michalis Michael[6], Rogiros Tapakis[6], and George Boultadakis[7]

[1] Research Institute of Science and Technology, Loyola University Andalusia, Seville, Spain
meliassi@uloyola.es

[2] Research Center on Renewable Electrical Energy Systems, Technical University of Catalonia, Barcelona, Spain

[3] Centro de Investigação em Energia REN - State Grid, S.A, Sacavém, Portugal

[4] Institute of Communications and Computer Systems, 9 Iroon Polytechniou Street, 15780 Athens, Greece

[5] Independent Power Transmission Operator (IPTO TSO), Athens, Greece

[6] Transmission System Operator, Cyprus (TSOC), Strovolos, Cyprus

[7] European Dynamics Luxembourg SA., 12, Rue Jean Engling, 1466 Luxembourg City, Luxembourg

Abstract. Nowadays the interest in grid-supporting energy storage systems for frequency response improvement is spurred to increase the penetration of renewable energy resources. Operational frequency constraints of the grid code should be fulfilled in the combined state feedback frequency control provided through the BESS frequency support and UFLS relays. In this paper, favoritism and unfairness of the grid-interactive Battery Energy Storage System (BESS) frequency support is investigated in terms of Rate of Change of Frequency (RoCoF), frequency nadir, time response, steady-state error, and specifically, total load shed subject to power balance over the network. Categorizing load shedding stages into vital and non-vital can measure the appropriateness of the BESS response. Conflicts of the BESS control parameters, performance measures and UFLS actions are verified on the modified Cypriot transmission grid, and the simulation results show that a controller or modulation technique would be essential to coordinate the BESS and UFLS scheme frequency responses to handle conflict of controllers.

Keywords: Battery Energy Storage System · Under Frequency Load Shedding · Dynamic frequency response · Performance measures

1 Introduction

A rising interest to investigate the impact of high penetration of renewable generation on frequency control of power system and the capability of delivering frequency support by full Converter Control-Based Generators (CCBGs) is emerged during recent

B. Németh and L. Ekonomou (Eds.): ISH 2019, LNEE 610, pp. 61–72, 2020.
https://doi.org/10.1007/978-3-030-37818-9_6

years [1]. Decrease of system inertia and conventional spinning reserve and increase of unpredictable uncertainties imposed by integrating renewables in the generation mix deteriorates the frequency support. By implementation of the virtual synchronous machine behavior [2, 3] through the new grid-supporting convertors of storages and renewables, the frequency support can be improved by means of the provided fast synthetic inertia and timely generation control in severe conditions [4].

In severe frequency decline situations, Under-Frequency Load Shedding (UFLS) programs (generally including event-based and response-based UFLSs) as the last automated measure of power system prevention from collapsing are designed and implemented by Transmission System Operator (TSO) through dynamic off-line simulations in operational planning time-scale. In response-based UFLS programs, the relay settings are usually pre-defined, fixed, and non-robust which may not be a comprehensive solution for the wide range of combinational/cascading events and operational variability and uncertainty imposed by high penetration of renewables [5]. Grid-interactive BESS can support the system operation during inertial and primary response through the inertia and droop parameters tuning of the SPC (Synchronous Power Controller).

Several TSOs have announced new grid codes requiring Electronically Interfaced Resources (EIRs) like energy storages, PVs, HVDC links, and wind power plants to provide frequency response [6]. The mutual interaction of UFLS as a traditional emergency control action has not been investigated enough under high share of renewable energy sources with BESS frequency support. The influence of the fast acting power controller of BESS on UFLS scheme and also mal-operation and readjustment of UFLS settings in high penetration of wind generation are presented in [7] and [8]. In [9], a systematic method is presented for controlling an energy storage system output power for preventing transient load shedding. Existing UFLS schemes suggested for conventional power systems has not been yet incorporated in grid-supporting convertors tuning to provide better inertial and primary frequency response.

A SPC-based Battery Energy Storage System (BESS) [10–12] as a linear state feedback controller could be tuned to control the performance of system frequency response measured by conflicting multi-dimensional performance measures such as minimum RoCoF and frequency, steady state error, and settling time. Good performance of UFLS schemes as another state feedback controller in conjunction with BESS frequency response is of great importance considering the fact that frequency support of the BESS may be harmful to some frequency performance measures or UFLS scheme performance. On the other hand, UFLS actions as a step change in RHS of the swing equation declines the BESS frequency response. Although, UFLS scheme has similar operation horizon as BESS frequency support, they used to be adjusted in different operational planning horizons, which makes the conflicts more complex and inevitable. In this paper, it is shown that the gains of the SPC and UFLS relays parameters need be adjusted in a coordinated manner to compromise between diverse technical and economic performance criteria of the frequency response.

The rest of the paper is constructed as follows: Sects. 2 and 3 provide an investigation of the impact of UFLS scheme and BESS inertia and droop characteristic on frequency response based on the unified discretized system frequency response using UFLS and BESS frequency responses formulation. Simulation results and discussion is presented in Sect. 4, respectively. The conclusion is drawn in Sect. 5.

2 Impact of BESS Inertia and Droop Characteristic on Frequency Response

2.1 Unified Discretized System Frequency Response

Based on the swing equation, accelerating and decelerating behaviour of the power system could be studied, when a power disturbance occurs. If the sum of all torques acting on the rotor shaft of a synchronous machine on the right-hand side of the swing equation including;

- the mechanical torque (Tm) (power input),
- the electrical torque (Te) (power output), and
- the damping torque (Td) [Nm]

Does not add up to zero, the excess torque accelerates or decelerates the rotor with moment of inertia J [kg m^2], meaning that the rotor angle θ [rad] accelerates (non-zero second derivative). By multiplying the swing equation with the system frequency ωb, the torques turn into power. In a single-area be made of Ng synchronous machine, the generalized continuous form of Center of Inertia (CoI) swing equation, presented in (1) is linearized as (2).

$$\frac{2H}{f_0}\frac{df(t)}{dt} = \sum\nolimits_{i=1}^{N_g} P_{mi} - P_{ei} \tag{1}$$

$$\frac{d\Delta f(t)}{dt} = \frac{f_0}{2H}\Delta P^{im}(t) \tag{2}$$

$$\Delta P^{im}(t) = \Delta P^{gov}(t) - \Delta P^{c}(t) + \Delta P^{sh}(t) + \Delta P^{wind}(t) + \Delta P^{PV}(t) + \Delta P^{ESS}(t) + \Delta P^{DG}(t) + \Delta P^{Tie}(t) - D\Delta f(t)$$

Terms of $\Delta P^{im}(t)$ could be defined as a function of $\frac{d\Delta f(t)}{dt}$ and/or $\Delta f(t)$ or as a step input. Generation and load changes which may cause input power imbalance are: Generation outage (ΔP^{c}), Governor action (ΔP^{gov}), Wind, PV, and ESS frequency response $(\Delta P^{wind}, \Delta P^{PV}(t), \Delta P^{ESS}(t))$ and load changes is incurred by load shedding $(\Delta P^{sh}(t))$, and load damping $(D\Delta f(t))$. By assuming each term of ΔP^{im} as X, the system frequency response could be discretized over time with time step Δt as $\Delta X(n\Delta t) = \Delta X_n$.

$$\Delta f_{n+1} = \Delta f_n + \Delta t\frac{f_0}{2H}(\Delta P_n^{gov} - \Delta P^{c} + \Delta P_n^{sh} + \Delta P_n^{wind} + \Delta P_n^{PV} + \Delta P_n^{ESS} - D\Delta f_n) \tag{3}$$

$$\Delta P_{n+1}^{gov} = \Delta P_n^{gov} + \frac{-\Delta t}{T}\left(\Delta P_n^{gov} + \frac{\Delta f_n}{R}\right) \tag{4}$$

The main focus of the rest of the paper is on the BESS influence on system frequency response and UFLS scheme. According to, other terms of (3) except than $\Delta P^{ESS}(t)$ and $\Delta P^{sh}(t)$ are ignored in the following equations. However, in simulation step, all of the resources contributing in the system frequency response are modelled and reflected.

2.2 System Acceleration Behaviour

The corresponding performance measures of the system frequency response are shown in Fig. 1.

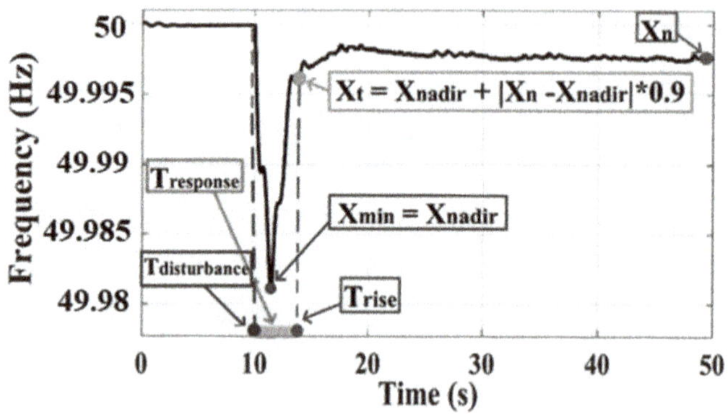

Fig. 1. Frequency performance measures illustration.

For a step-wise decrease in ΔP^c from 0 to $-\Delta P$ at the disturbance time t_0, system frequency starts to drop fast with a high RoCoF. Higher RoCoF may activate the RoCoF relays of DGs and loads. The inertial and primary frequency response of the generations and loads try to limit the frequency deviation by increasing the system kinetic energy. At minimum frequency (Nadir frequency) time t_m, RoCoF reaches zero and the frequency deviation Δf starts to decrease. Subsequently, Δf oscillates and finally stabilizes at the new steady-state frequency Δf_{ss}.

Consequently, as shown in Fig. 2 the frequency response could be divided into areas in which the power system is accelerating, and others in which the system is decelerating. In an accelerating area, the frequency deviation Δf with respect to the steady-state frequency Δf_{ss} is increasing or in other words $\Delta f - \Delta f_{ss}$ and RoCoF have the same sign. This can be written as:

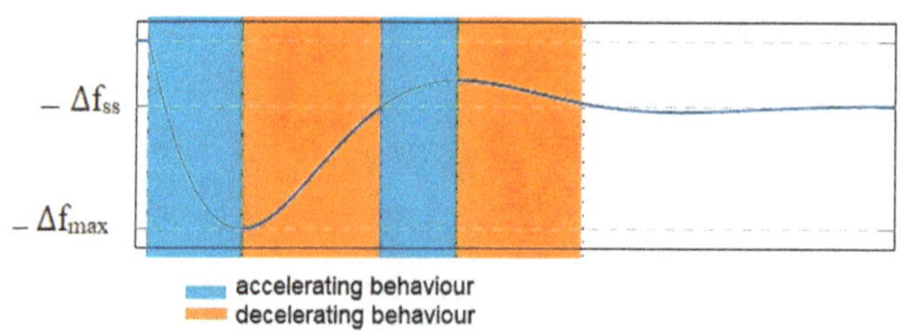

Fig. 2. Acceleration periods of power system frequency response.

$$\text{Accelerating} : (\Delta f(t) - \Delta f_{ss}).\frac{d\Delta f(t)}{dt} > 0 \tag{5}$$

$$\text{Decelerating} : (\Delta f(t) - \Delta f_{ss}).\frac{d\Delta f(t)}{dt} < 0$$

2.3 BESS as a State Feedback Controller

The frequency response of storage can be similar to conventional units when participating in inertia and primary response. When the system is faced with large amount of power deficiency, ESS can provide FFR (Fast Frequency Response) based on the following transfer function [13]:

$$\Delta f(t) = \frac{\Delta P^{ESS}(t)}{K_e} \qquad K_e = \frac{-\left(2H_e s + \frac{1}{R_e}\right)}{1 + T_e s} \tag{6}$$

where H_e and R_e are inertia constant and primary frequency constant of control, respectively, and T_e is the time constant of BESS which is ignored in comparison to the time constants of the other and consistent with conventional units. Therefore;

$$-2H_e\frac{d\Delta f(t)}{dt} - \frac{1}{R_e}\Delta f(t) = \Delta P^{ESS}(t) \tag{7}$$

$$-2H_e\frac{\Delta f_{n+1} - \Delta f_n}{\Delta t} - \frac{1}{R_e}\Delta f_n = \Delta P^{ESS}(n) \tag{8}$$

Charging, discharging and power capacity limits of the BESS is assumed to be considered on the limits of the ΔP^{ESS}, H_e and R_e. Integration of (8) in (3) results in (9).

$$\Delta f_{n+1} = \Delta f_n + \Delta t\frac{f_0}{2(H + H_e)}(\Delta P_n^{gov} - \Delta P^c + \Delta P_n^{sh} - \frac{1}{R_e}\Delta f_n - D\Delta f_n) \tag{9}$$

In (9), BESS inertia and droop characteristics turn to be a state feedback on system frequency. In order to provide a better frequency response, we need to tune the inertia and damping of the BESS, which is achieved by regulating the converters' SPC

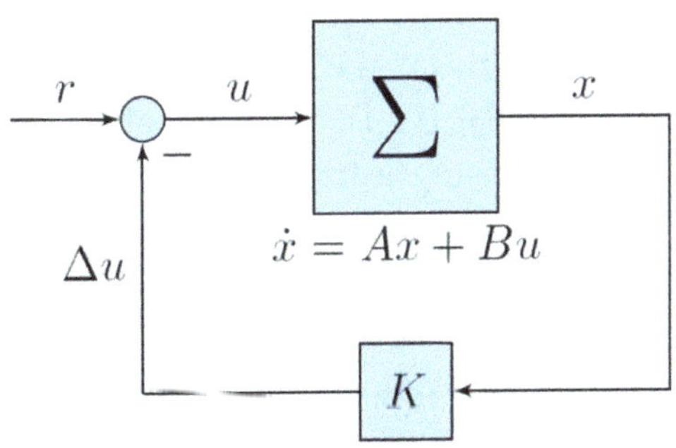

Fig. 3. SPC-based BESS as a State feedback control.

parameters (H and R) through a Multi-objective optimal gain-tuning method. A SPC-based BESS as a state feedback of the system DAEs is shown in Fig. 3 in which x is $[f, \dot{f}]$ and K = [H, R] of the controller.

2.4 Discussion on BESS Frequency Response

We may state that feedback can cause a system that is originally stable to become unstable. Certainly, feedback is a double-edged sword; when it is improperly used, it can be harmful [14]. Based on the (5) and (6), during accelerating time-intervals, to arrest the rate of change of frequency the system needs inertia emulated by the BESS. As soon as the system starts decelerating, inertia should reduce again to prevent overshoot in frequency response. The unused inertia imposes the frequency to return to the steady-state frequency slower, which deteriorates the overall frequency response.

Additionally, damping should be increased during decelerating phases to help decrease the settling time of the frequency deviation. The additional inertia required during accelerating phases is proportional to the RoCoF and the required additional damping is proportional to the frequency deviation from the steady state. Without adaptive damping and inertia control, the BESS response is not canceled out when the system has sufficiently stabilized which may result in unstable oscillations.

Due to the conflicting influence of the battery inertia and droop responses in accelerating and decelerating periods of frequency response, conflict of interests appears between different performance measures of frequency response with respect to the provided inertia and droop responses of BESS.

3 Response-Driven UFLS as a Non-linear State Feedback Controller

The response driven based load shedding, widely uses Under Frequency-based Load Shedding solutions. The inherent closed loop/ feedback based control scheme in decentralized response driven UFLS system makes them efficient in acting against the disturbances.

As frequency is a very good indicator of Power Mismatch, the operation based on under frequency relays has high control precision and robustness with respect to uncertainties. However, it can have many difficulties and challenges. The setting of the UFLS is highly complex, which involves many parameters including the number of LS stages, percentage of load allowed to be shed, the time delay for each stage, real time topology etc. Together these variables make the UFLS settings a non-linear and multi-dimensional problem. Secondly, UFLS is triggered after the frequency has already declined to certain low values causing a time delay, which makes the solution more reactive in nature and the system stability requires more time.

UFLS relays are characterized using incremented/decremented step behavior in swing equation. For simplicity, the related blocks can be represented as a sum of incremental (decremental) step functions. For instance, as presented in (10), for a fixed UFLS scheme [15], the function of $\Delta P^{sh}(t)$ in the time domain could be considered as a sum of the incremental step functions of $\Delta P_k.u(t - t_k)$. Therefore, for L load shedding steps:

$$\Delta P^{sh}(n) = \sum\nolimits_{k=1}^{L} \Delta P_k.u(n - n_k) \tag{10}$$

$$u(n - n_k) = 1 \text{ if } f_0 + \Delta f_{n-n_k+i} < f_{th}^k \quad i = 0, \ldots, n_k \tag{11}$$

(10) and (11) could be linearized using binary and big variables [16]. As stated above, the relay logic dictates that the block of load ΔP_k be shed when the corresponding timer exceeds n_k. When a contingency occurs, for each load-shedding stage k with f_{th}^k, the relay disconnects a block of load after n_k time step when the frequency trajectory [computed using (3)–(4)] violates the frequency set point for a predetermined time delay.

3.1 Discussion on UFLS and BESS Mutual Effect

Similar to the SPC-based BESS frequency response, a response-driven based UFLS scheme is affecting the frequency response as a non-linear state feedback controller as presented in Fig. 4. **f(x)** is presented in (10).

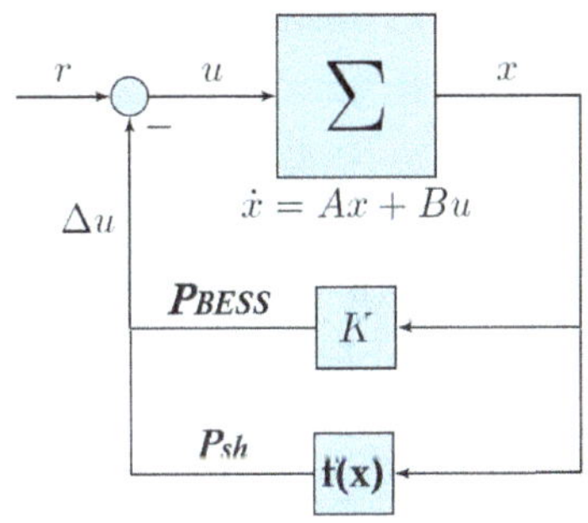

Fig. 4. SPC-based BESS and UFLS scheme as State feedback controls control.

UFLS action at $t_{f_{th}} - t_k$ is a positive step change in generation-load imbalance which is reflected in system DAEs as a disturbance. Subsequent change in the rocof and frequency values alter the BESS inertial and primary response and reduce the BESS frequency support.

In some situations, however, frequency support of EIRs may be harmful to frequency control as they generate an extra energy in a short period of time, which reduces frequency decay and derivative but could not be maintained over time. Reduction of frequency derivative through emulating inertia via EIRs looks positive at first sight as the reduced frequency derivative triggers less under-frequency relays, but in some conditions, the shedded load is smaller than the amount what the event requires. As the EIRs cannot maintain extra generation over time, the frequency may continue to decay until the shedding of the next load step. It is also possible to activate other system and components protection by resulting in lower steady state frequency.

Based on (10) and (11), load shedding steps are activated, if the frequency reduces temporary or recurrently below a certain UFLS threshold. If the steady state frequency settles at a frequency above the same threshold, load shedding may be considered as *nonvital*. BESS can provide fast injection of power with limited energy capacity to prevent temporary frequency decreases and unnecessary load shedding. On the other hand, BESS should withdraw from support when the frequency decline is under arrest and additional load shedding would not be activated.

In order to avoid mal operation of UFLS scheme in the presence of the BESS frequency support, UFLS schemes should be incorporated in the controller tuning of the BESS as an economic objective to be optimized in conjunction with other frequency performance measures. In comparison with BESS controller tuning, most power systems operate with predetermined load shedding schemes and use fixed relay settings. Hence, resetting UFLS thresholds or changing load shedding amounts are not often viable options.

Moreover, with the existing supervisory and control capabilities, TSOs are not tendentious to re-design UFLS scheme for a short while. Then, the frequency thresholds, time delays and shed load percentage could be assumed fixed during the optimal gain-tuning of BESS. Using synchrophasor measurements for control of grid interactive energy storage system and UFLS relays [17] provide the possibility of coordinated online control of these resources of inertia and primary frequency response.

4 Studied System

4.1 System Modelling and Characteristics

There are 13 load shedding stages ($\sim$68% of the total load) spread over all Cyprus, which are presented in the following Table 1. This UFLS scheme is implemented in the system by definition and setting of the under frequency relays in DIgSILENT PowerFactory. A 50 MW, 100MWh BESS is added to the network equipped by SPC with tunable inertia and droop parameters of 15 and 60 as the maximum gains, as it is presented in [18]. The initial values of 0.5 and 0 are assumed for the SOC and power output states of the BESS, respectively. SOC limits are supposed as 0.2 and 0.8.

Other RHS terms of the swing equation in (2) are enclosed in the system modelling and control like turbine and governors, and Wind turbine standard controllers accorded to Grid Code requirements. Load damping effect are not considered in this study. The coincident disturbance of two generation units with 90 and 60 MW dispatch power is considered to tune the BESS parameters.

Table 1. Load shedding stages spread over all Cyprus.

Stages	Frequency thresholds	Time delay	Shed Percentage	Shed MW
1	49	0.2	4	37.69
2	48.9	0.2	4	37.69
3	48.8	0.2	3	28.27
4	48.7	0.2	8	75.38
5	48.6	0.2	4	37.69
6	48.5	0.2	4	37.69
7	48.4	0.2	4	37.69
8	48.3	0.2	1	9.42
9	48.2	0.2	7	65.95
10	48.1	0.2	6	56.53
11	48	0.2	5	47.11
12	47.75	0.2	9	84.80
13	47.5	0.2	9	84.80

4.2 Conflict of Interests

In order to investigate the mutual effect of UFLS plan and BESS frequency support, system frequency response is presented in different scenarios in Fig. 5. As it is shown, without UFLS relays and BESS (a), with UFLS relays and without BESS (b), and with BESS (H = 5, R = 20) and UFLS (c) scenarios are studied.

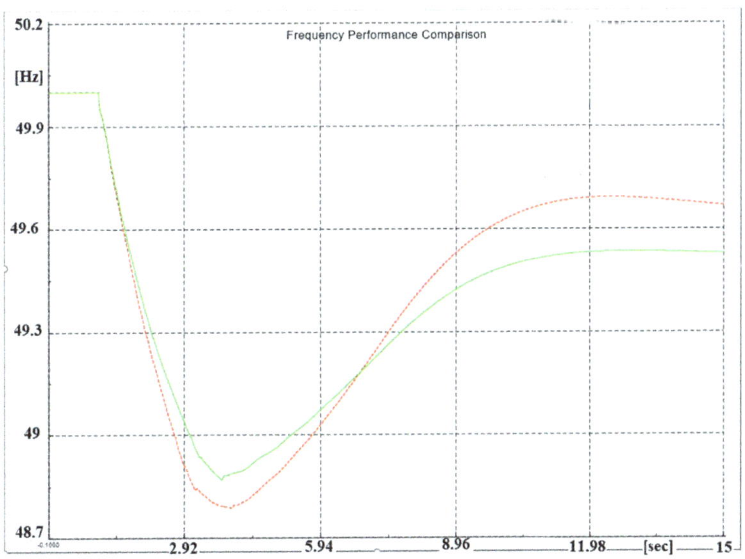

Fig. 5. Effect of UFLS plan and BESS frequency support on frequency response: red (b) and green (c).

Comparison of transient responses in these scenarios are summarized as follows:

1. In (a), the system frequency decreases continuously until the system becomes unstable due to the cascading outage of conventional and renewable resources due to their under frequency relay activation. Primary and inertia support of conventional generation are not enough to arrest the frequency decline.
2. By activation of UFLS in (b), the first three blocks of the UFLS plan are activated with 0.2 s delay after the frequency reaches the thresholds (t_{action} = [2.89, 3.15, 3.95]). The non-vital shedding stage is avoided by BESS frequency support. Reproducing the frequency responses for three different combination of activated shedding stages in Fig. 6 (red: stage one, green: stage one and two, blue: all stages) analyses the necessity of the activated stage with respect to steady state frequency. While two first stages are categorized as vital, third stage is non-vital as the frequency settles above 48.8 without the activation of this stage.

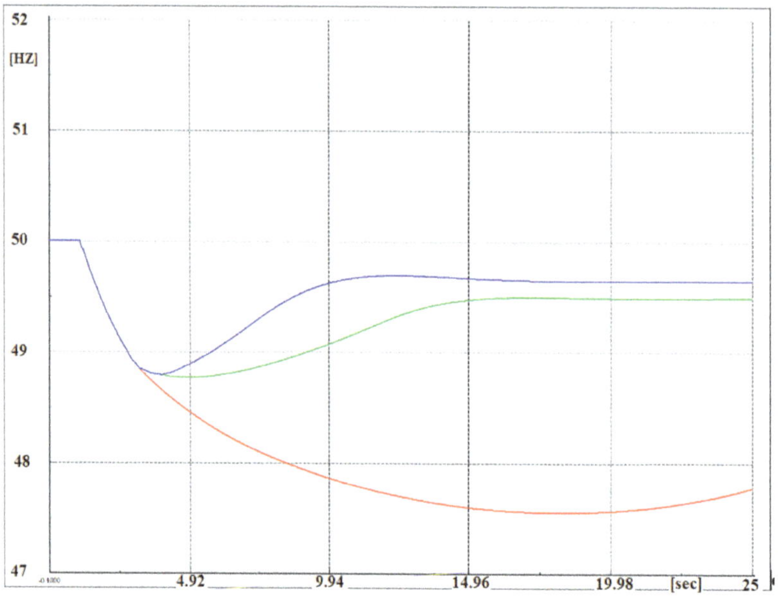

Fig. 6. Vitality analysis of activated load shedding stages.

3. Addition of BESS in (c), the first two blocks of the UFLS plan are activated with 0.2 s delay after the frequency reaches the thresholds (t_{action} = [3.24, 3.74]). Although the non-vital shedding stage is avoided by BESS frequency support, other performance measures like RoCoF and minimum frequency are improved and response time and steady state frequency affected harmfully.

As it is presented in Table 2, a conflict of interests appears due to the parallel activation of UFLS and BESS frequency support in viewpoint of different performance measures. Undesirable effect of the BESS are grayed out in the table.

Table 2. Frequency performance measures.

Performance measures	With BESS	Without BESS
f_{Nadir}	48.87	48.78
$RoCoF_{min}$	−0.0845	−0.0859
F_{ss}	49.53	49.67
Response time	8.57	8.44
t_{Nadir}	3.74	3.96
P_{shed}	75.34	103.48

5 Conclusion

First, it was shown that by considering BESS droop and inertia gains along with the multistage UFLS plan, the overall performance of the load shedding plan and BESS frequency support are mutually influenced by each other in the both favorable and unfair directions. It was also revealed that the BESS droop and inertia gains are affecting frequency performance measures as a double-edged sword regarding system acceleration behaviour. Accordingly, the conflict of interests raised with respect to the UFLS and BESS state feedback controls interference (with similar operation and different planning horizons) should be handled through a gain tuning approach. Enabling coordinated online control of BESS parameters UFLS scheme using synchrophasor measurements provides better inertia and primary frequency response considering demonstrated conflicts of interest.

Acknowledgments. This work was partially supported by the European Commission under project FLEXITRANSTORE—H2020-LCE-2016-2017-SGS-774407 and by the Spanish Ministry of Science under project ENE2017-88889-C2-1-R. Any opinions, findings and conclusions or recommendations expressed in this material are those of the authors and do not necessarily reflect those of the host institutions or funders.

References

1. Dreidy, M., Mokhlis, H., Mekhilef, S.: Inertia response and frequency control techniques for renewable energy sources: a review. Renew. Sustain. Energy Rev. **69**(November 2015), 144–155 (2017)
2. Rodriguez, P., Citro, C., Candela, J.I., Rocabert, J., Luna, A.: Flexible grid connection and islanding of SPC-based PV power converters. IEEE Trans. Ind. Appl. **54**(3), 2690–2702 (2018)
3. D'Arco, S., Suul, J.A., Fosso, O.B.: A virtual synchronous machine implementation for distributed control of power converters in SmartGrids. Electr. Power Syst. Res. **122**, 180–197 (2015)
4. Morren, J., de Haan, S.W.H., Kling, W.L., Ferreira, J.A.: Wind turbines emulating inertia and supporting primary frequency control. IEEE Trans. Power Syst. **21**(1), 433–434 (2006)
5. Tang, L., McCalley, J.: Two-stage load control for severe under-frequency conditions. IEEE Trans. Power Syst. **31**(3), 1943–1953 (2016)

6. Greenwood, D.M., Lim, K.Y., Patsios, C., Lyons, P.F., Lim, Y.S., Taylor, P.C.: Frequency response services designed for energy storage. Appl. Energy **203**, 115–127 (2017)
7. Gonzalez-Longatt, F., Rueda, J., Vázquez Martínez, E.: Effect of fast acting power controller of battery energy storage systems in the under-frequency load shedding scheme. In: International Conference on Innovative Smart Grid Technologies (ISGT Asia 2018), no. June (2018)
8. Aparicio, N., Añó-Villalba, S., Belenguer, E., Blasco-Gimenez, R.: Automatic under-frequency load shedding mal-operation in power systems with high wind power penetration. Math. Comput. Simul. **146**(2017), 200–209 (2018)
9. Pulendran, S., Tate, J.E.: Energy storage system control for prevention of transient under-frequency load shedding, pp. 1–11 (2015)
10. Zhang, W., Cantarellas, A.M., Rocabert, J., Luna, A., Rodriguez, P.: Synchronous power controller with flexible droop characteristics for renewable power generation systems. IEEE Trans. Sustain. Energy **7**(4), 1572–1582 (2016)
11. Marin, L., Tarras, A., Candela, I., Rodriguez, P.: Stability analysis of a grid-connected VSC controlled by SPC
12. Eliassi, M., Paulino, P.A.B., Torkzadeh, R., Rodriguez, P.: Event-based under-frequency inertia emulation scheme for severe conditions
13. Khazaei, J., Nguyen, D.H., Thao, N.G.M.: Primary and secondary voltage/frequency controller design for energy storage devices using consensus theory. In: 2017 6th International Conference on Renewable Energy Research and Applications, ICRERA 2017, vol. 2017–Janua (2017)
14. Kuo, F.G.B.: Automatic Control Systems, pp. 398–401 (2002)
15. Hassan Bevrani, B., Verbic, G.S., Koepper, H.D.: Robust Power System Frequency Control, vol. 48, no. 2 (2009)
16. Banijamali, S., Amraee, T.: Semi adaptive setting of under frequency load shedding relays considering credible generation outage scenarios. IEEE Trans. Power Deliv. (2018)
17. Torkzadeh, R., Eliassi, M., Mazidi, P., Rodriguez, P.: Synchrophasor measurements for control of grid interactive energy storage system
18. Rocabert, J., Luna, A., Blaabjerg, F.: Control of power converters in AC microgrids. IEEE Trans. Power Electron. **27**(11), 4734–4749 (2012)

Power System Studies in the Clean Energy Era: From Capacity to Flexibility Adequacy Through Research and Innovation

Vasiliki Vita[1(✉)], Elias Zafiropoulos[1], Ioannis F. Gonos[1], Valeri Mladenov[2], and Veselin Chobanov[2]

[1] Institute of Communications and Computer Systems, 9 Iroon Polytechniou Street, 15780 Athens, Greece
vasvita@aspete.gr

[2] Technical University of Sofia, 8 KL. Ochridski Blvd., 1000 Sofia, Bulgaria

Abstract. A secure and reliable supply of electricity is a necessary requirement of a well-functioning economy. Reducing the likelihood of system disturbances or, in the extreme, load (demand) shedding is a fundamental objective of Transmission System Operators during planning and operation of electricity grids. The integration of new forms of generation - diverse in location and operation profile – as well as rapidly changing demand from new consumer technologies powered by breakthroughs in technology, poses a different set of challenges in their daily business. System flexibility, like many other aspects of the bulk electric system, can be examined as a supply and demand problem. A holistic assessment of flexibility examines the resources available to supply flexibility and the factors driving the demand or requirements for flexibility. This paper aims to highlight the importance of flexibility assessment studies alongside capacity adequacy studies being carried out during system planning. Moreover, the paper will present the challenges facing the electricity networks for integrating high amount of renewable energy which are well expressed by the views of the energy stakeholders regarding priorities for innovation technologies and this is also very well reflected in the research and development projects. The FLEXITRANSTORE project, which is aiming to implement innovative technologies of battery storage solutions, smart grid technologies and market platforms in the SEE region, is briefly presented. Furthermore, a wide review of flexibility-related research and innovation projects in Europe and United States has been carried out during the project and their contribution is presented, analysed and mapped against specific power system challenges. A detailed survey conducted during FLEXITRANSTORE on identifying stakeholders' opinion on electricity networks' challenges is presented and commented. Stakeholders' are stressing that highest priorities for the electricity networks in their country are renewable energy integration, cross border interconnections, grid stability in short- and long-term horizon. These renewables integration and cross border interconnections answers could be also linked with the European targets: To sum up, the paper is identifying significant opportunities for future research and innovation in storage and market models, regarding the inherent flexibility of the electricity networks to host the electrification 'wave' of the following decades.

B. Németh and L. Ekonomou (Eds.): ISH 2019, LNEE 610, pp. 73–83, 2020.
https://doi.org/10.1007/978-3-030-37818-9_7

Keywords: Flexibility · FLEXITRANSTORE project · Power systems · Transmission systems operators

1 Introduction

Planning and operation of modern electric power systems comprehend several complex and interlinked tasks. These tasks can be divided in three main groups, depending on the considered time horizon: (i) Long-term resource and equipment planning, which targets time ranges from one year to several decades. Examples are investment planning, transmission and distribution planning and long range fuel planning, (ii) Short-term operational scheduling, which is used for time intervals from several hours to a few weeks, or even year(s). Examples are unit commitment (UC) scheduling, maintenance and production scheduling and fuel scheduling, (iii) Real time operations, which consider fractions of a second to several minutes [1, 2].

Concern for the environment and energy security, as well as rising fuel prices, have led to significant, sustained growth of wind and solar electricity generation capacity in Europe and worldwide. The difficulty posed by the integration of these variable resources into existing power systems varies according to the production and scale of the variable resource, its correlation with system load, and the flexibility of the power system in question. Flexibility can be defined as the ability of a system to deploy its resources to respond to changes in net load, where net load is defined as the remaining system load not served by variable generation [3–5]. The impact of RES on the different time scales is depicted in Fig. 1.

Capacity adequacy studies have addressed the question of how much capacity is required to reliably meet system load, at a certain point in time, but have not considered whether the system's planned resources could be operated in a sufficiently flexible manner. These studies are key tasks for regulatory bodies, who look to ensure that market designs deliver in the long run, or for vertically integrated utilities, ensuring the suitability of a planned plant portfolio before more detailed engineering and operational analyses are carried out. Investors and plant manufacturers also have an interest in the cycling requirements experienced by potential investments. Variable renewables (VRES) integration studies have also required enhanced simulation and modeling tools. The adaptation of unit commitment to a stochastic [6] and rolling framework, and the inclusion of VRES forecasts [7] can provide an insight into how systems might operate under high a variable generation penetration.

This paper aims to highlight the importance of flexibility assessment studies alongside capacity adequacy studies being carried out during system planning. The challenges facing the electricity networks for integrating high amount of renewable energy are well reflected in the views of the energy stakeholders regarding priorities for innovation technologies and this is also very well reflected in the R&D projects. To sum up, the paper is identifying opportunities for future research and innovation regarding the inherent flexibility of the electricity networks to host the electrification 'wave' of the following decades.

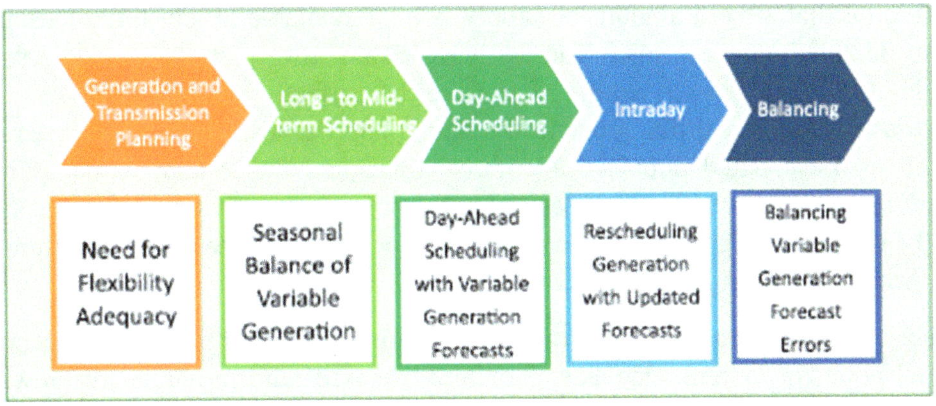

Fig. 1. Impacts of intermittent RES on different timescales of power system operation [3, 4].

2 The Need for Flexibility Assessment

Since the construction of a new generation facility has a multi-year lead time, traditional long-term planning is required to ensure the future reliability of power systems. The Capacity Adequacy Forecast (CAF) has been assessing the capacity adequacy of available supply resources to meet simulated demand scenarios. This annual exercise provides an overview of the state-of-play across a ten-year time frame, carried out by national TSOs for each country and ENTSOe for the European network [8]. Ten years is usually the time frame during which policy makers, investors and market participants make strategic decisions so as to allow the power system to deliver a targeted level of adequacy. Such forecasting exercises focus on the likelihood of unusual events disrupting supply. The results are derived from many simulations providing a probabilistic interpretation of the likelihood of lack of supply. One such metric explored in the CAF and covered here, is loss-of-load expectation (LOLE). Simply put, it is the average number of hours per year in which it is statistically expected that there is not sufficient power supply in the market to cover demand. Note that this is not translated in a blackout as the analysis keeps security margin for unforeseen events near real time, since it is a probabilistic assessment based on models.

Adequacy is not only related to the total amount of capacity being installed in the system, but also to the ability of the installed capacity to adjust to the ever-increasing dynamics of dispatch events in the system. The latter is defined as flexibility adequacy and it becomes ever more important, mainly due to the increasing amount of variable renewable energy present in the power system [1, 5]. Several flexibility services will be required in order to ensure a smooth transition to high RES penetration, particularly:

- Ramping needs: With the high RES penetration, flexible resources will become essential to meet the fast change of residual demand i.e. the steep upward ramp created by the decline in solar output due to the sun setting when demand increases in the evening.
- Balancing fast reserves: The increase of variable generation along with the forecast error of wind and PV should be overcome with reserve deployment in order to secure the supply. Modelling of balancing reserves in the CAF is performed

assuming that a fixed amount of supply is kept available at any time. Despite the considerable improvements in forecasting variable power generation, for both wind and solar, in practice, forecasting can never be perfectly accurate, with decreasing forecast errors as real-time operation moment approaches. Thus, forecasts are very likely to be updated hours ahead of real time, and the system will require fast starting and controllable resources (interconnectors, demand side response, storage and fast response generators). It is easy to imagine that the larger the ramps the bigger the need for flexibility.

Recent studies in Europe and around the world confirm that flexibility is becoming a crucial point for system adequacy. Flexibility services and products are growing in importance and are progressively being integrated into the market. The ramping needs already play a central role in US electricity markets and flexibility services are being designed and implemented in European markets. This tendency towards flexibility services is more than evident in the Research & Innovation efforts in Europe and US [2, 9, 10].

3 Flexibility Research and Innovation in the Horizon 2020 Context: The FLEXITRANSTORE Project

Identifying the aforementioned challenges for, several European R&D projects and Innovation Actions have been working intensely on the issues of integrating renewable energy technology into the electricity grids. Looking into the European and US projects related with RES integration and flexibility issues, as these have been identified in [11], interesting remarks may come out:

(a) Extensive applications of Distributed Energy Resources (DER) integration are being developed and implemented in real life scenarios. They focus mostly on the balancing and congestion challenges, voltage/frequency regulation and stability, communications infrastructure, smart energy management, micro-generation and distributed storage. There are very few applications studying market solutions for integrating DER through providing flexibility services and remuneration mechanisms.

(b) Among the various flexibility related EU projects, the distribution grids, local electricity networks and microgrids have been very popular for technology demonstrations. Large scale technology solutions in transmission networks have been demonstrated in few European projects, focusing either on improving grid assets efficiency and promoting HVDC technology (BESTPATHS), or on developing simulation platforms on market, real time system operation or cross-border trading (FUTUREFLOW, CROSSBOW). Flexibility assessment frameworks for transmission networks, bridging technology innovations, system simulation methodology and market reforms proposals, have not been the main driver in EU research programs so far.

(c) Storage related European projects have been targeting mainly to the LV-MV level i.e., battery storage in building blocks, residential thermal storage or local RES-storage. Thus, the related flexibility services have been demonstrated at the distribution level with no clear plan for upscaling to the transmission system
(d) On the other side of the Atlantic, research and innovation projects in the US are closely related to the new technologies developed by the power equipment manufacturers. Selective products are chosen to be demonstrated in projects to promote industrial innovation and focus greatly on developing new business models and revenue streams. That is to say, while promoting grid operation benefits for flexibility, the majority of projects depend greatly on the respective business prospects.
(e) Energy storage plays a key role in the modernization of the US power grid [10]. Several storage projects with variable scales have been implemented in the US electricity networks, while the US market regimes promote the provision and remuneration of storage related grid services, such as frequency regulation, power balancing, black start capability, smoothing, load shifting, ramp management, energy price arbitrage. An example is the PJM RTO, which structures its frequency regulation market so that energy storage is compensated for the speed and accuracy it can provide to the system [9].
(f) These energy storage projects include a variety of RES-plus-storage applications. Significant large scale projects of wind generation coordinated with battery storage have been commissioned in the US, such as the Tehachapi project [12]. The wind-storage demonstrates the aforementioned services, as well as further support in the respective area of Southern California with high wind capacity: load shed deferral for increasing system reliability of supply, renewable-energy-related transmission optimization, transmission congestion issues mitigation.
(g) Electric Power Research Institute (EPRI) has been developing a system flexibility assessment tool InFLEXIon, currently version 3.0. This software tool was developed in the context of a research program led by EPRI and can assist planners in assessing operational flexibility of electric power systems, analysing system requirements for supply-demand energy balancing, supporting long-term generation and transmission planning [13].
(h) Additionally, the International Energy Association (IEA) has developed FAST and FAST2 tools in the context of "The Grid Integration of Variable Renewables (GIVAR)" undertaken by the IEA to address the critical question of how to balance power systems featuring large shares of VRE. Several countries worldwide were studied and significant results were presented in the respective reports [14].

The aforementioned review was conducted in the context of an ongoing Innovation Action Project FLEXITRANSTORE and the results of the survey are mapped in Table 1. The project FLEXITRANSTORE has launched its activities in November 2017 and will focus on demonstrating transmission grid innovative technologies that improve flexibility in the SEE power system area.

Grid scale storage at the HV/MV substations and respective efficient controllers will be demonstrated for two different applications: TSO-DSO coordination, depicting

Table 1. Flexibility related EU-funded projects and their topics of interest

PROJECT ACRONYM \ Objectives/ Research Areas	Renewable integration	Energy Storage	Fast/Flexible dispatchable generation	Smart asset management	Flexible ac transmission systems	Demand response	New grid infrastructure reinforcement	Power quality	Grid stability	Cross border interconnections	Ancillary services remuneration	Wholesale market	Retail Market /prosumer	Dynamic pricing	TSO/DSO coordination	Market coupling
SMARTNET		√								√		√		√	√	
ELSA		√	√				√	√	√							
FLEX4GRID							√						√			√
EMPOWER			√				√						√			√
ENERGISE						√	√									
UPGRID			√			√			√	√			√			
MIGRATE	√		√	√	√	√	√	√								
TILOS	√		√						√							
TRIANGULUM				√		√	√									√
SMARTER EMC2	√					√			√							√
IDE4L		√		√		√	√	√	√				√	√		
GRID+STORAGE		√													√	√
VENTEEA	√						√	√								
NETFFICIENT	√	√	√	√			√		√				√			
NAIADES		√	√													
STORY		√		√			√									√
FLEXMETER				√		√	√		√				√			√
REALVALUE		√	√				√					√				
SENSIBLE	√	√		√			√		√							√
P2P SMARTTEST		√					√		√							√
STORE&GO	√	√														
NOBEL GRID				√			√						√			
BESTPATHS				√		√	√	√							√	
INDUSTRE	√					√					√	√	√	√		√
FLEXICIENCY							√				√	√	√			√
FUTUREFLOW			√			√			√	√		√				√
DAREED	√					√	√		√							√
DISCERN	√					√	√	√	√							
eBADGE			√			√		√	√							√
ANYPLACE		√		√		√	√					√	√	√		
ELECTRA	√					√			√	√		√			√	√
INSPIRE-GRID	√						√									√
HYBRID-VPP4DSO	√		√	√		√	√		√							
BESTRES	√	√	√			√					√	√	√			√
INTEGRID	√					√		√	√		√	√				
PROMOTION	√				√		√	√		√						
EVOLVDSO	√		√				√		√					√	√	√
INTERFLEX	√	√		√		√							√			
GRID4EU	√	√		√		√		√					√			

demand side management flexibility potential and Wind-Storage configuration, promoting ancillary services of a 'near-dispatchable' nature. Business models and market platforms will be investigated and demonstrated respectively, and solutions will be proposed, especially for demand side management flexibility. The benefits of these projects will be related with flexibility adequacy studies in Greece, Bulgaria and Cyprus. In Table 1, brief statistics show that most popular topics are new equipment on the grid, demand response, RES integration, market coupling and grid stability. However, Cross border cooperation and TSO-DSO coordination need further attention by R&I community.

4 Stakeholders' Opinion on Flexibility Challenges: Where We Are Today and Future Prospects-Aspirations

In FLEXITRANSTORE, a questionnaire was launched to stakeholders asking their opinion in various aspects of flexibility as well as their interest on the specific technology innovations promoted in this project. During an almost 3-month period, 47 answers were received across Europe and across sectors including network operators, regulators and policy makers, market operators, ESCOs, aggregators, energy suppliers, power producers, manufacturers and research community.

The first 7 questions of the questionnaire cover a variety of issues on power system flexibility and electricity markets, giving the stakeholders the opportunity to express their opinions and limitations they face in their national and pan-European electricity sector. The answers provided excellent information on the current international technologies and assessment frameworks that are related with operational flexibility, and identifying the key concerns and misperceptions of the term. All answers stressed the importance of flexibility, beyond operational reliability, presenting examples of inflexibility in technical level and market level, emphasizing on the gaps and needs for flexibility assessment.

In the question 8, asking for the critical topics for the power system and electricity markets in a short and long term time horizon, with a rate of relevance, where 1 = low relevance and 5 = high relevance, the stakeholders had the chance to outline the power system challenges in a qualitative way. The outcomes were averaged and sorted accordingly in two separate diagrams shown side by side in Figs. 2 and 3. Furthermore, the answers of value 4 or 5 (meaning high relevance) have been allocated to the critical topics and the corresponding percentages are presented in Figs. 4 and 5.

According to stakeholders, highest priorities for the electricity networks in their country are renewable energy integration, cross border interconnections, grid stability in short and long term horizon. RES integration and cross border interconnections answers could be also linked with the European targets. Moreover, in short term horizon (up to 5 years) the wholesale markets and the ancillary services remuneration (market–related) are highly prioritized as critical topics. In the long term horizon (more than 5 years), energy storage and demand response are rising higher to the prioritization list of critical topics, since they are currently considered rather immature in Europe. Therefore, significant research and innovation are needed on these topics:

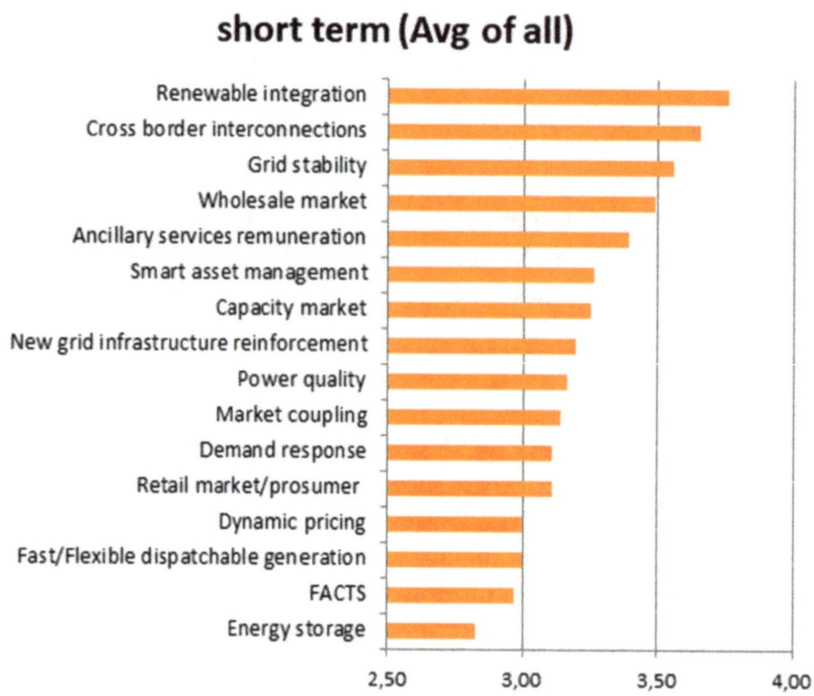

Fig. 2. Average value of relevance, according to stakeholders, on critical topics related to flexibility, in short (1–5 years) horizon.

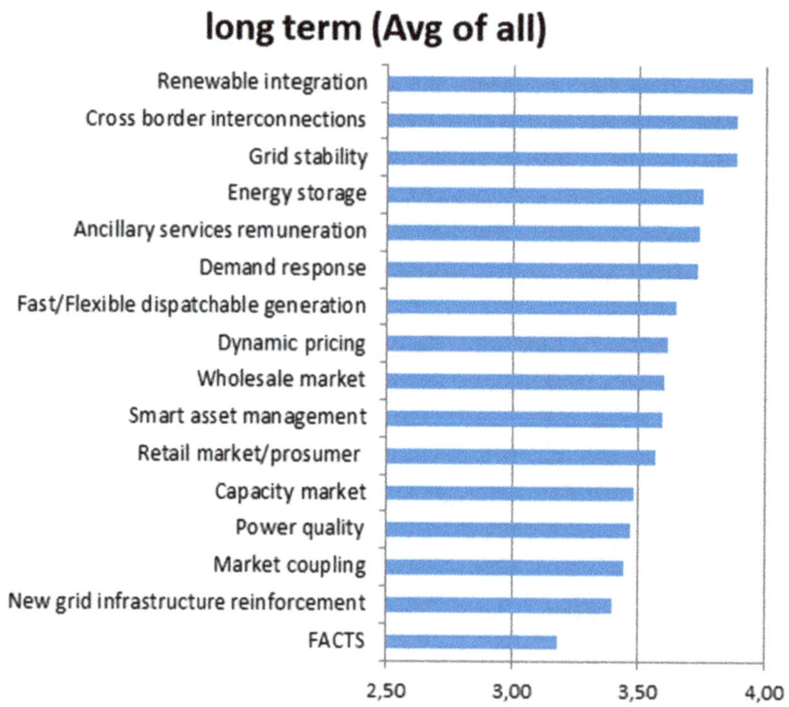

Fig. 3. Average value of relevance, according to stakeholders, on critical topics related to flexibility, in long (6–10 years) horizon.

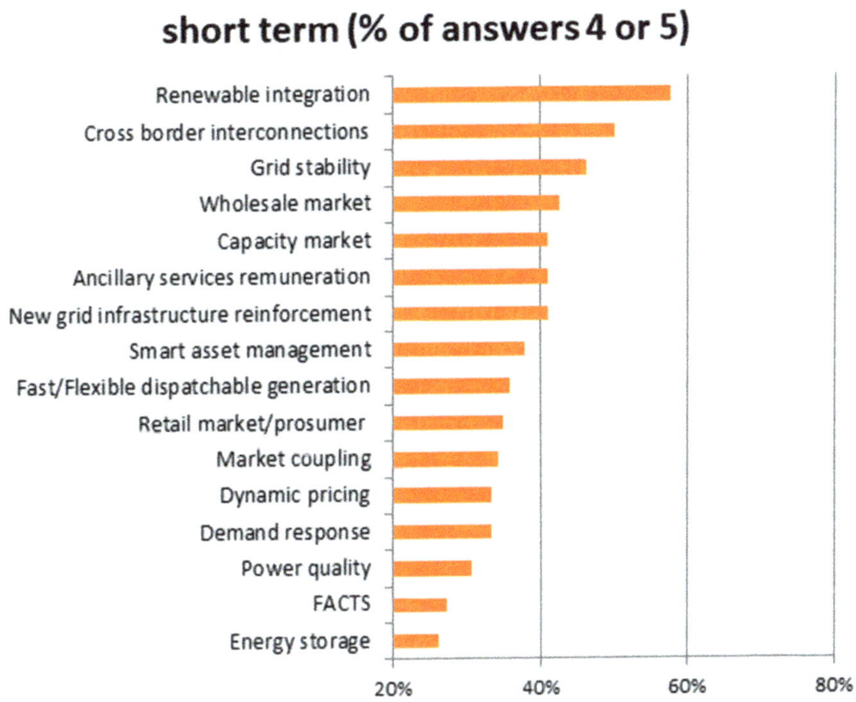

Fig. 4. Percentage of number with high relevance, according to stakeholders, over total answers on critical topics related to flexibility, in short (1–5 years) horizon.

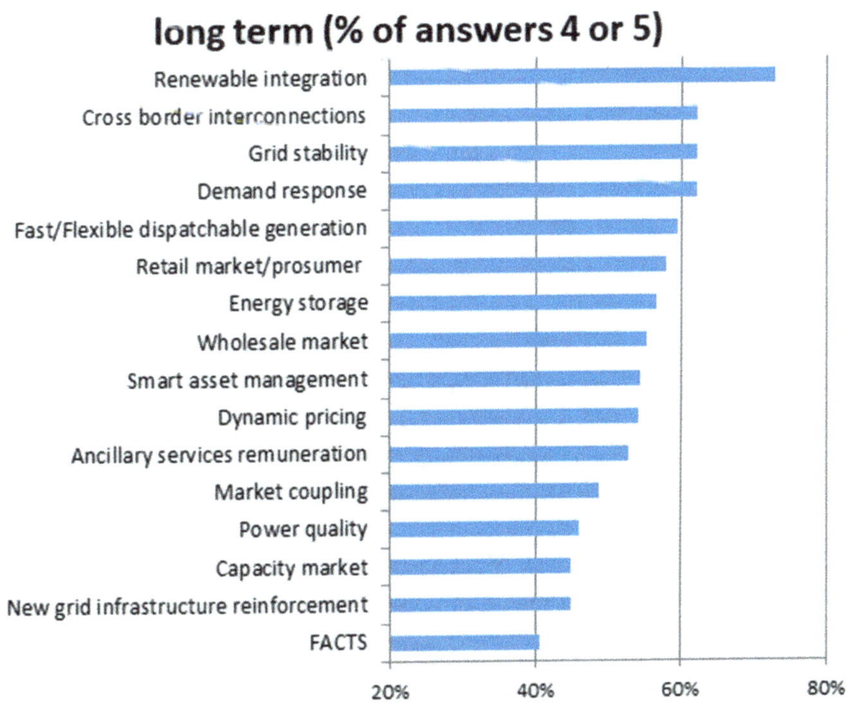

Fig. 5. Percentage of number with high relevance, according to stakeholders, over total answers on critical topics related to flexibility, in long (6–10 years) horizon.

(i) EU countries have already agreed on a new renewable energy target of at least 27% of final energy consumption in the EU as a whole by 2030 as part of the EU's energy and climate goals for 2030.
(ii) The October 2014 European Council called for interconnection of at least 10% of installed electricity production in the Member States by 2020, endorsed the 15% target by 2030 and underlined that they will be both attained via implementation of Projects of Common Interest in energy infrastructure.

Moreover, in short term horizon (up to 5 years) the wholesale markets and the ancillary services remuneration (market related) are highly prioritized as critical topics. In the long term horizon (more than 5 years), energy storage and demand response are rising higher to the prioritization list of critical topics, since they are currently considered rather immature in Europe.

5 Conclusions

Power systems will be called upon to host a wide share of RES generation, bringing the flexibility assessment for needs and resources to the forefront. In the conclusions of the questionnaire by FLEXITRANSTORE project, stakeholders are stressing greatly on renewable integration and enhanced interconnections, while they are greatly interested in energy storage and market incentives for investments and innovation. Wholesale markets with flexibility services remuneration are also of great interest among stakeholders. EU-funded project FLEXITRANSTORE is focusing on demonstrating smart grid and market solutions for increasing transmission grid flexibility joining the efforts of academics, technology providers, industry, market and networks operators.

Acknowledgment. This project has received funding from the European Union's Horizon 2020 research and innovation programme under grant agreement No. 774407.

References

1. Ulbig, A.: Operational flexibility in electric power systems. Ph.D. dissertation EEH, ETH, Zurich, Switzerland, vol. 72, no. 21882, 230 (2014)
2. Fraunhofer-Institute for Wind Energy and Energy System Technology (IWES): The European power system in 2030: Flexibility challenges and integration benefits, pp. 1–88. Agora Energiewende (2015)
3. Holttinen, H., Tuohy, A., Milligan, M., Lannoye, E., Silva, V., Muller, S., Soder, L.: The flexibility workout: managing variable resources and assessing the need for power system modification. IEEE Power Energy Mag. **11**(6), 53–62 (2013)
4. Ecofys: Flexibility options in electricity systems; Project number: POWDE14426, 10 March 2014. www.ecofys.com
5. Lannoye, E., Flynn, D., O'Malley, M.: Transmission, variable generation, and power system flexibility. IEEE Trans. Power Syst. **30**(1), 57–66 (2015)
6. Wu, L., Shahidehpour, M., et al.: Stochastic security-constrained unit commitment. IEEE Trans. Power Syst. **22**(2), 800–811 (2007)

7. Tuohy, A., Meibom, P., et al.: Unit commitment for systems with significant wind penetration. IEEE Trans. Power Syst. **24**(2), 592–601 (2009)
8. ENTSOe: MAF 2018 – Mid-term Adequacy Forecast (2018)
9. Chen, H., Baker, S., Benner, S., Berner, A., Liu, J.: PJM integrates energy storage: their technologies and wholesale products. IEEE Power Energy Mag. **15**(5), 59–67 (2017)
10. Fyer, J., Corey, G.: Energy storage for the electricity grid: benefits and market potential assessment guide. Sandia National Laboratories, SAND2010-0815 (2010)
11. Sustainable Places Conference. http://www.sustainableplaces.eu/
12. Gaillac, L., et al.: Tehachapi Wind Energy Storage Project: Description of operational uses, system components, and testing plans. In: IEEE PES T&D 2012, Orlando, FL, USA (2012)
13. www.epri.com
14. www.iea.org

Battery Energy Storage System Integration in a Combined Cycle Power Plant for the Purpose of the Angular and Voltage Stability

François Kremer[1,2(✉)], Dominique Remy[1], Wangue Merville[1], Stéphane Rael[2], and Matthieu Urbain[2]

[1] General Electric, Rue de la découverte, 90000 Belfort, France
francois.kremer@ge.com
[2] GREEN, Université de Lorraine, 54500 Vandoeuvre lès Nancy, France

Abstract. The introduction of the renewable energy sources in the electrical network is something necessary in the ecological transition context. This integration forces the grid operators to change the way to generate electrical power by favoring sustainable sources against traditional techniques. That means, adapt the grid code to provide the same energy quality for the end-customer despite the growing grid instability. Fast acting systems become more and more necessary to stabilize the grid by limiting the frequency, power and voltage swings. The integration of an electrochemical storage system at different connection points of the network is a solution technically studied over the last decade. In this article, the BESS will be connected inside a powerplant with a focus on the hybridization of the sources. Services as the angular stability and the plant voltage regulation are described however services concerning frequency regulation or plant design improvement are not approached. The hybridization concerns the synchronous machine, the associated excitation, the storage and the control system. The expected benefits for the grid operators are the enhancement of the flexibility, the robustness and the electrical quality towards a more and more constraining grid instability.

Different services will be presented describing the integration concept, the potential connections, the sizing or some operating profiles. These services will concern the small-signal stability, the transient angular stability, the voltage regulation and the voltage disturbances (flicker, harmonics…). A simulation has been done with Matlab/Simulink™ to show the benefits of the solution under the technical requirement specification n°6 of the French grid operator.

Keywords: BESS · Thermal power plant · Angular stability · Voltage stability

1 Introduction

The market of electricity is today undergoing change due to the introduction of renewable power sources. These integrations and environmental challenges result in strong destabilization of the electricity grid. It can be noticed a loss of traditional generation groups leading to an increasing active and reactive power variability. A way

B. Németh and L. Ekonomou (Eds.): ISH 2019, LNEE 610, pp. 84–94, 2020.
https://doi.org/10.1007/978-3-030-37818-9_8

to fight against theses disturbances is to improve the flexibility of the network. This is defined by Belgian TSO as "the capacity of the system to react towards un-predictable and predictable grid changes while ensuring the security, the reliability and the efficiency of it" [1]. A mix of flexible sources, as generation units (Combined Cycle Gas Turbine, biomass, hydraulic, nuclear …), energy storage systems and a predictive management of the demand at the Point of Connection (PoC) are necessary to ensure the stability of the electricity network [2].

Many articles explain the benefits of BESS connected at different points of the grid and whose profitability is demonstrated thanks to grid ancillary services [3–6]. Ancillary services being the technical capacity of the system to meet the optional or mandatory requirements imposed by the TSO to flexibility sources in the grid code. These requirements are related to the frequency, the voltage and the angular stability or the grid restauration in case of failure (black start). The ancillary services provided by the BESS have the disadvantage to have a limited time duration due to the limited energy reservoir. Therefore, the historical way of generation like CCGT (Combined Cycle Gas Turbine) or hydropower stations are mainly used to perform theses grid services. These sources offer a continuous grid support despite their low reactivity due to mechanical time constants.

A stable quality of service should be maintained, defined as the capacity to stay and return at a stable operating point, with the ancillary services, which are retributed and structured in 3 categories:

The angular stability: The synchronous generator is connected to the grid and rotate at the synchronous speed due to an elastic torque created by the electrical values [2]. Some events could disturb this synchronization leading to a trip of the production units. The angular stability is the capacity to be and stay connected to the grid. In other words, to maintain the internal load angle of the machine within certain limits depending on the active and reactive power. The de-synchronization, which is nothing more than a loss of control of the internal angle, can be caused by severe load variations called transient stability or small disturbances, called small-signal stability [7].

The voltage stability: The grid equipment is defined for one single voltage value, and the variation outside the nominal operating range involves material destruction, early ageing, additional losses, … This variation can be induced by the imbalance between the power consumption and the power generation (like the frequency regulation) or some changes of the network topology. To reduce the imbalance, a primary, secondary and tertiary reserves are used [3].

The frequency stability: The frequency stability aim is to maintain the balance between the power generation and the power demand, different control loops have been implemented for this purpose. These regulators avoid, first, the frequency collapse (frequency containment) then re-establish the frequency to its initial value (frequency restoration), and finally restore manually the reserve (replacement reserve).

This document proposes an improvement of the grid flexibility by combining two complementary sources of flexibility which are a thermal power plant (CCGT) and the battery energy storage system. The powerplant reference chosen in the article does not impact the way to use the BESS as long as the services focus on the angular and voltage stability. Indeed, the equipment concerned by these functions are the synchronous machine and its excitation system, the storage system and the control which are comparable for different powerplants. Only the single line diagram or the PoC might be modified.

The hybridization of a BESS and a combined cycle power plant is not achieved yet. Indeed, the West Burton power plant integrates a Li-ion storage system to provide EFR. However, no software integration has been performed to hybridize the gas turbine and the BESS. Another example can be the hybridization of an aero-derivative gas turbine in Norwalk, California with a BESS by General Electric. The system can provide frequency response, voltage support, spinning reserve and blackstart. In this example, the power ratio between the BESS and the gas turbine is important (20%) compared to our application in which the ratio will be less than 5%. The hybrid solution presented in this paper proposes a complete control integration, a battery optimization size and known services, but provided by both systems.

After a brief description of each system, services achievable by the hybridization are explained in detailed with different topologies and control philosophies. Finally, a hybrid service is simulated under the French grid codes rules.

2 Hybrid System

2.1 Combined Cycles Gas Turbine

The CCGTs are composed by a Gas Turbine (GT) associated with a Steam Turbine (ST). The concept of this plant is to improve the overall Power Plant efficiency by using the hot exhaust gas to boil water in a Heat Recovery Steam Generator producing steam

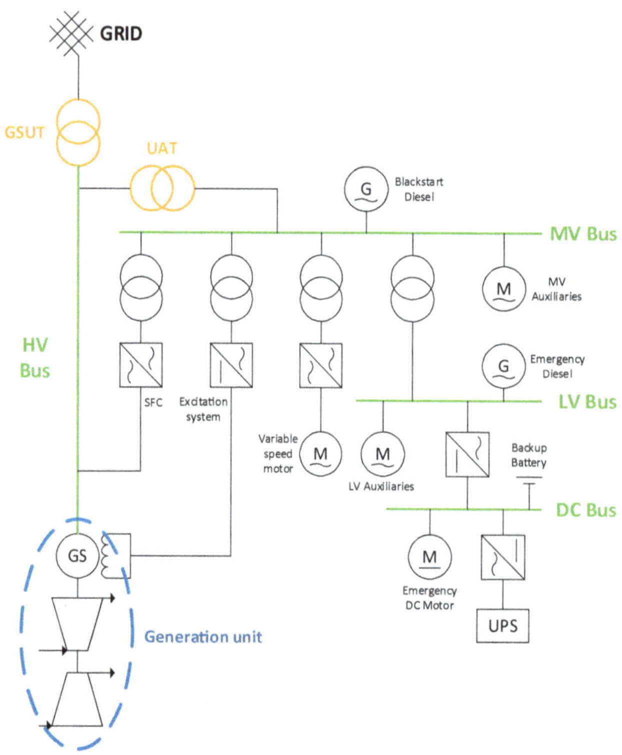

Fig. 1. Basic Single Line Diagram of CCGT with gas turbine auxiliaries

to feed the ST. The electricity is produced with synchronous generators driven by the turbine(s) shaft(s). One can approximate that, the power delivered by the ST is roughly the half of whole GT (Fig. 1).

On the electrical side, the generator is directly connected to the grid by means of a HV bus called the Insulated Phase Bus and the Generator Step-Up Transformer (GSUT). All auxiliaries as pumps, motors, water tanks or inverters are fed via the Unit Auxiliary Transformer (UAT) and the auxiliary buses (MV, LV or DC Bus) connected to the HV.

2.2 Battery Energy Storage System

Concerning the electrochemical storage, the lithium technology is chosen for its high cycling capacity and for its fast charge and discharge possibilities. To manage this electrical flux, the BESS needs power electronics to drive, to convert and to regulate the transferred energy. The other main elements of the storage device are the BMS (Battery Management System), the MV/LV or HV/LV transformer depending of the point of connection and the controllers which shall manage set-point signal, communication and dispatch instruction interfaces for battery, converter and grid system.

The integration of a BESS into the CCGT in term of layout, connection and control shall be designed carefully as it has an influence on plant equipment. Difficulties is even worse when the BESS is added to an existing plant.

3 Angular and Voltage Stability Services

Depending on the point of connection of the BESS, the provided services are different. A stand-alone BESS due to its location will improve either the distribution or transmission grid whereas the PI-BESS will improve the distribution services and the plant operability.

A non-exhaustive list of services is presented in a paper under acceptance, the services discussed here focus only on the angular and voltage stability as shown in the next table. Their aims are to enhance the operability of the generator by avoiding the disconnection or by supporting the auxiliary bus (Table 1).

Table 1. Services concerning angular or voltage stability

Services names	
QE1	Transient angular stability
QE2	Small-signal stability
QE3	Voltage regulation
QE4	Disturbances reduction

3.1 QE1: Transient Angular Stability

The transient angular stability is a deviation of the internal load angle of the machine due to a severe and temporary disturbance. Criteria considered is a voltage dip or a sudden fall of the voltage greater than 10%. The consequence of the voltage dip is the reduction of the resistant electrical torque. Without brake, the generator will accelerate up to the default suppression threshold and then suddenly decelerate trying to stabilize the rotational speed. Sudden accelerations of the generator shaft will increase the internal load angle with the risk to desynchronize the synchronous machine from the grid and force the generator to disconnect. In that case the synchronous torque between the grid and the generator must be maintained by the excitation system by boosting or damping the flux in the generator air-gap.

In case of a stator voltage dip, the PI-BESS can improve the angular stability by providing or by absorbing an instantaneous additional reactive or active power. The principle is the equal area criterion which aim is to have a "net-zero" energy balance [7].

The integration of such a control needs to operate in combination with the excitation, without bringing power oscillation between the two different regulators. In the case of new units, a centralized and common controller should be implemented to manage both systems at the same time to propose better robustness of the whole system. In the case of powerplant retrofits, the communication cannot be easily implemented. The decentralized control, which aim is to maintain the global stabilization by locally support the grid, will be the adequate solution to use. For instance, General Electric has implemented decentralized control with an aeroderivative turbine and a BESS in California.

In any case, the PI-BESS power supply will be limited by the converter size and a voltage regulation loop must be implemented to support locally the generator and increase the LVRT (Low Voltage Ride Through) compliance. The point of connection will be either on the MV bus close to the UAT with limited power or on the HV bus to obtain a better impact on the transient angular stability (Fig. 2).

Another way to integrate the PI-BESS, is to replace the excitation system by the BESS, with DC/DC converter between the exciter and the battery. In that case, the conventional excitation bridge (one way) is substituted by a bidirectional AC/DC converter to charge and discharge the battery and an additional DC/DC converter to provide the DC field current to the generator. This configuration could allow the stabilization by providing at the same time, an over boost of the excitation and a damping power with the inverter. This deep integration will contribute to have a better response of the entire system but will force to change a lot of equipment and usage, which cannot be easily provided on plant retrofit (Fig. 3).

[3] estimates a high-power BESS, around 10% of the generator nominal power to get a significant impact on the distribution lines. However, a hybrid solution with a capacity of 2% of the generator nominal power, for the 1st integration case (retrofit) is impacting as shown in the last section. No study has been achieved for the second case. The mechanism of valorization does not exist yet and the economic profitability for the angular stability cannot be considered. It is the primary difficulty during a commercial bid, to translate this technical benefit in economic value.

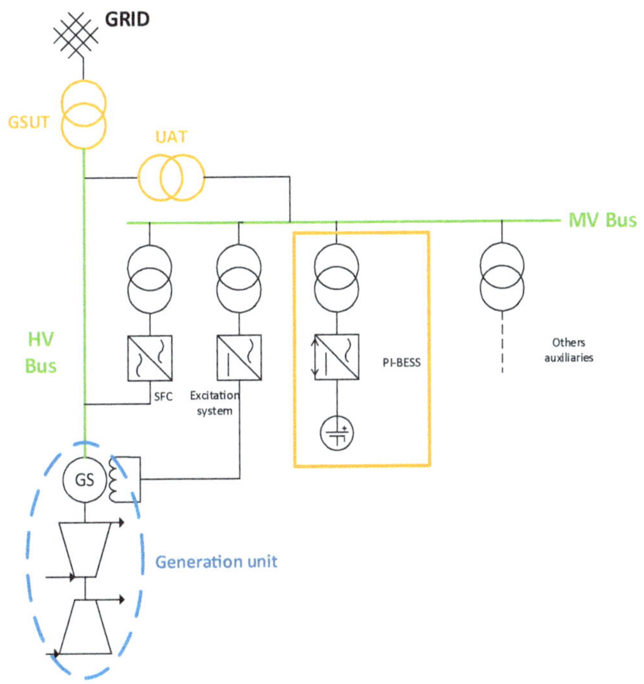

Fig. 2. Integration of the BESS in parallel to the excitation system.

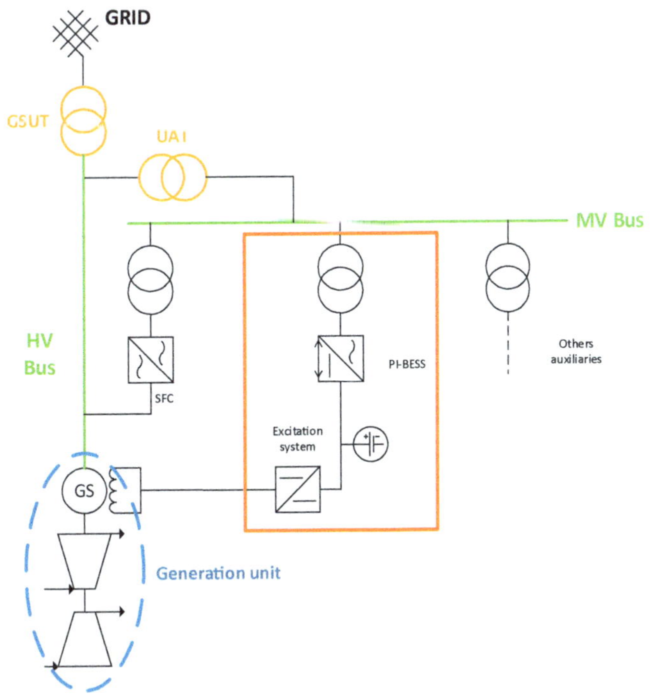

Fig. 3. Integration of the BESS in series with excitation system.

3.2 QE2: Small-Signal Stability

Unlike the transient angular regulation, the small-signal stability must be insured continuously and is induced by small oscillating instabilities (0.2 to 2 Hz) around the operating point. These oscillations, apparent on the voltage, are normally smoothed by the damping torque of the synchronous machine. [8] explains that the modern excitation systems usually increase the synchronism torque (smoothing the severe disturbances) however it reduces the damping torque. In doing so, the machine is more sensitive to small-signal disturbances. A supplementary control layer, called Power System Stabilizer (PSS), is generally added to the Automatic Voltage Regulator (AVR) to deal with the inter-area oscillations using a damping torque.

The damping torque can be produced, as explained in the previous section, with a control loop of the excitation system (PSS) or with FACTS. The FACTS systems (Flexible AC Transmission System) are used to improve the voltage quality on transmission lines by providing only reactive power with power electronics devices. Similarly, to the FACTS, the PSS loop could be implemented in the PI-BESS to generate the appropriate power to smooth the rotor load angle oscillations.

The use of this service must be carefully adapted to the location. Indeed, a wrong location can amplify the small signal disturbances. To avoid this, fast communications between the excitation system and the storage system are justified. The power needed for small signal disturbances is unknown. Especially if the PSS control loop is dispatched both in the excitation system and the BESS, which will highly complexify the stability study.

As the service QE1, the storage system can be integrated either on the HV bus or the MV bus depending the amount of the power transfer needed. Further study will be driven to size the storage system and study the control stability. The benefit of the solution cannot be remunerated but proposed for a better energy quality.

3.3 QE3: Voltage Regulation

As is it usually the case in the transmission grid, the voltage regulation is based only on reactive power control for inductive load. The regulator will control the amount of reactive power and the response time. In the case of PI-BESS, the reactive power is essentially provided by the generator due to the power limitation of the converter characteristics. The reactive power is produced with phase shift between the current and the voltage on AC side. The battery energy used will only compensate the loss in the converter and globally stay at the same state of charge (SoC). It is the reason why the limiting factor for this service is the maximum converter power and not the stored energy.

The battery will support the local MV area (i.e.: voltage dip caused by pump start) or help the transient response of the synchronous machine by providing faster reactive power (<100 ms). Indeed, the small power of the BESS compared to the synchronous generator MVArs limits its effect on the generator stability. The PI-BESS can be used as a boost to complement the excitation system, as shown Fig. 3. This topology will bring better and faster control than the actual technology without problem of voltage drop. The replacement of the actual robust and controlled excitation system by a new system with battery is useless without merging battery services.

[3] proposes a sizing method for the converter around few MVARs but it has to be considered that the cost of the solution is not competitive compared to the passive solutions (capacitors banks, tap changer, AVR in the synchronous machine) for distribution lines. In our case, the passive solutions are not relevant since the excitation system itself is enough to solve this issue. The integrated asset will focus on all aspects of the quality of energy by doing multiple services and not only QE3.

The used of the storage system and the way to connect and control it is like QE1–QE2 services. In that case, the magnitude can be different depending on the wished effect on the plant. Indeed, 20% of rated generator power is needed to impact the overall plant but 2% is enough to support pump start on the MV bus.

The added value of this service is limited since the synchronous generator is designed to provide or absorb an important volume of MVARs. However, the additional MVARs can be an extra-service without costing more during the sizing.

3.4 QE4: Disturbances Reduction

The stability of the electrical grid depends on the generation units, the transmission and distribution operators and the consumers. Indeed, disturbances can be created at each level of the grid. This energy quality is assessed with electrical factor such as the voltage continuity (long or short power interruptions) and the quality of the voltage (amplitude or frequency variation, distortions or unbalances). At local level, the generation assets help for the voltage continuity, but their main usage is to maintain the quality of the grid according to the applicable grid code. In France, EN50160 and IEC61000 are the standards which define criteria for normal operation conditions for the voltage quality.

The disturbances which deteriorate the voltage quality are the flicker, the voltage drops, the harmonics, the overvoltage and the voltage phases unbalances. The usual methods (passive filters, AVR and PSS, capacitor banks, …) to compensate the disturbances can be replaced by power electronics or in that case a BESS. For instance, in the case of harmonic distortions the converter will be used as an active shunt compensator injecting the adequate harmonic current. As every consumer on the grid, the auxiliary bus has the duty to manage the quality of the voltage. With the connection on the transport grid, the short circuit power is large enough to insure a good voltage quality. However, on islanding mode (weak grid), the local installation needs to have harmonic filters to provide the correct operation due to a small short circuit power. The PI-BESS is a solution to regulate the power quality.

The benefit of this service cannot be directly financially estimated but needs to be translated in an improvement of the grid operation (losses reduction) or a decrease in grid quality financial penalties. An investment shift can be another benefit, using the asset instead of installing passive solutions.

The BESS sizing depends on the point of connection and the level of disturbances to damp, it will be in the order of few MVA for few minutes of use. This hybrid solution can be used for blackstart service which is usually ensure with diesel generator and harmonics filter. The challenge will be to size and control correctly the storage system.

4 Sizing and Benefit of the Solution

The benefit of such a system is difficult to understand and to evaluate without a retribution from the TSO. The interest of the BESS support can come to the spotlight with the increasing penetration of renewable energy, asking for more stringent Grid Code requirement. To demonstrate the interest to implement this solution, a simulation on the transient angular stability has been achieved demonstrating the technical benefit of such a solution.

The French grid operator RTE imposes different technical requirement specification which need to be respected by the installation connected on the transport grid (>50 kV) [9]. The sixth specification concerns the dynamic voltage regulation and the transient angular stability. After a robustness evaluation with margin calculation on different operating points, a SMIB (Single Machine – Infinite Bus) model is tested for a variation of 2% on the voltage set-point injected in the excitation system. This model needs to be tested with two different values of the grid impedance and pass the compliance criteria (response time, overshoot, …). The assessment of the excitation system had been achieved under these conditions.

The model is built with the Simscape library on MATLAB-SIMULINK. The synchronous machine, the three-Phase Transformer (Two Windings) and the Three-Phase Series RLC Branch models are the Simscape models. A small-signal low-frequency linear circuit model is implemented for the inverter. It models the average behavior of the inverter and the control loops (Power, Current and DC Voltage) using Concordia and Park transformation as explained in [3] (Fig. 54). The battery is supposed as a constant voltage source due to the small simulation time (few seconds). Finally, the excitation system is model by a ST4B presented in the IEEE 421.5. The simulation is validated with the 6th specification of the French grid code.

Considering the overall controller philosophy, no control loop of it has been considered in this paper. The reference of the storage system controller is immediately the power variation measured on the generator shaft. The detailed design of the control loop will follow in a further article.

The parameter use for the simulation are taken from a 2 × 1 7HA.02 powerplant of General Electric loaded at 65%. The short circuit power is around 10 000 MVA with a lagging power factor at 0.9 pu, a grid voltage at 95% and a generator voltage around 103% of the nominal value.

To prove the benefit of the PI-BESS, the behavior is checked without the storage system, with active power control and with active and reactive power control. The test consists of adding immediately 20 MVAr on the grid side. Without hybridization, the generator voltage reaches 105% and the angle have 2 degrees of deviation. The power control of the battery allows to smooth the transient variation of the angle. However, the 6 MW deliver cannot change the voltage deviation. Using the battery with active and reactive control, the variability of the electrical variables is damped. Indeed, the angle deviation is only 1 degree and the generator voltage reaches 104%. The reactive power profile is a step of 8MVAr absorb by the BESS.

These logical results prove an effect of the battery for the grid stabilization and allow to estimate the active and reactive power involved. With an adequate control, this benefit can improve the size of the battery and help the plant operator to meet stringent grid requirements (Fig. 4).

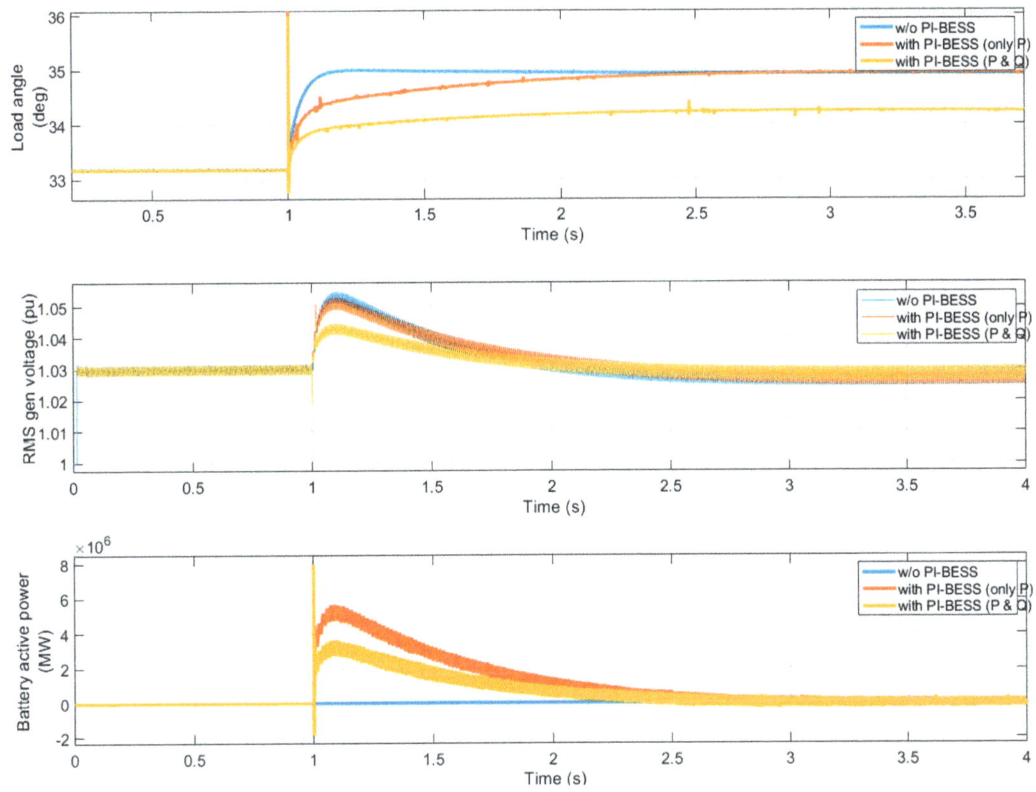

Fig. 4. Synchronous generator load angle, HV bus voltage and battery power variation under a 20 MVAr load shedding

5 Conclusions

Enhancements of the grid voltage quality and angular stability with an integration of BESS in a combined cycle are proposed in this article. These improvements concern the transient angular stability, the small-signal stability, the local voltage regulation and disturbances attenuation. The PI-BESS integration inside a power plant induce hardware and software development which will be plant specific and will be fine-tuned to avoid instability in case of large grid disturbances. As currently, there is no remuneration associated to the functionalities proposed by the article, we have decided to perform the first set of study via computer modelling only considering the French grid code specifications. Operation with and without PI-BESS has been shown. The first simulation results allow to validate these concepts and to estimate the appropriate capacity of the battery that would be needed to reach a significant impact on the generator dynamic services.

Acknowledgment. This paper is part of the FLEXITRANSTORE project. This project has received funding from the European Union's Horizon 2020 research and innovation program under grant agreement No. 77440.

References

1. Electricity scenarios for Belgium towards 2050, ELIA, November 2017. http://www.elia.be/~/media/files/Elia/About-Elia/Studies/20171114_ELIA_4584_AdequacyScenario.pdf. Accessed 27 May 2019
2. Mémento de la sûreté du système électrique, RTE, Technical report, ISBN - n° 2-912440-13-0 (2004). www.rte-france.com. Accessed 27 May 2019
3. Delille, G.M.A.: Contribution du Stockage à la Gestion Avancée des Systèmes Electriques: approches Organisationnelles et Technico-économiques dans les Réseaux de Distribution. Sciences de l'ingénieur [physics], Ecole Centrale de Lille, Français, <NNT: 2010ECLI0016> (2010)
4. Fitzgerald, G., Mandel, J., Morris, J., Touati, H.: The Economics of Battery Energy Storage: How Multi-use, Customer-Sited Batteries Deliver the Most Services and Value to Customers and the Grid. Rocky Mountain Institute, September 2015. http://www.rmi.org/electricity_battery_value. Accessed 27 May 2019
5. Christiansen, C., Murray, B., Conway, G.: Energy storage study: a storage market review and recommendations for funding and knowledge sharing priorities. AECOM, July 2015. https://arena.gov.au/assets/2015/07/AECOM-Energy-Storage-Study.pdf. Accessed 27 May 2019
6. Sidhu, A.S., Pollitt, M.G., Anaya, K.L.: A social cost benefit analysis of grid-scale electrical energy storage projects: evaluating the smarter network storage project. University of Cambridge, EPRG Paper, May 2017. https://www.eprg.group.cam.ac.uk/wp-content/uploads/2017/06/1710-Text.pdf. Accessed 27 May 2019
7. Kundur, P.: Power System Stability and Control. Electric Power Research Institute, Power System Engineering Series. McGraw-Hill (1994). ISBN 0-07-035958-X
8. Alkhatib, H.: Etude de la stabilité aux petites perturbations dans les grands réseaux électriques: optimisation de la régulation par une méthode métaheuristique, Automatique/Robotique, Université Paul Cézanne - Aix-Marseille III, Français (2008)
9. Documentation technique de référence, RTE, p. 699 (2009). http://clients.rte-france.com/htm/fr/mediatheque/telecharge/reftech/16-10-2017_complet.pdf. Accessed 27 May 2019

Synchrophasor Based Monitoring System for Grid Interactive Energy Storage System Control

Roozbeh Torkzadeh[1(✉)], Mojtaba Eliassi[1], Peyman Mazidi[1], Pedro Rodriguez[1,2], Dalibor Brnobić[3], Konstantinos F. Krommydas[4], Akylas C. Stratigakos[4], Christos Dikeakos[5], Michalis Michael[6], Rogiros Tapakis[6], Vasiliki Vita[7], Elias Zafiropoulos[7], Ricardo Pastor[8], and George Boultadakis[9]

[1] Research Institute of Science and Technology, Loyola University Andalusia, Seville, Spain
rtorkzadeh@ieee.org

[2] Research Center on Renewable Electrical Energy Systems, Technical University of Catalonia, Barcelona, Spain

[3] Studio Elektronike Rijeka (STER d. o. o), Rijeka, Croatia

[4] Department of Electrical and Computer Engineering, University of Patras, Rion, Greece

[5] Independent Power Transmission Operator (IPTO TSO), Athens, Greece

[6] Transmission System Operator, Cyprus (TSOC), Strovolos, Cyprus

[7] Institute of Communications and Computer Systems, 9 Iroon Polytechniou Street, 15780 Athens, Greece

[8] Centro de Investigação em Energia REN - State Grid, S.A, Sacavém, Portugal

[9] European Dynamics Luxembourg SA., 12, Rue Jean Engling, 1466 Luxembourg, Luxembourg

Abstract. Energy Storage Systems installed at primary substations can be used by different participants of the power system for handling the emerging uncertainties caused by supply-side variations, demand-side flexibility and grid topology changes. This work presents a practical design for the monitoring system of the controller of the grid interactive energy storage system. This monitoring scheme takes advantage of synchrophasor measurements gathered by phasor measurement units of wide area measurement systems. The analysis of synchrophasor measurements provides real-time situational awareness over the status of the grid. Therefore, the integration of synchrophasor measurements into the control loop of GI-ESS will enable them to participate in power services of the flexibility market. Furthermore, the implementation of a basic power oscillation damping function as an example of power services using the proposed monitoring system is illustrated in this paper.

Keywords: Phasor measurement unit · Grid interactive battery energy storage system · Wide area control system

B. Németh and L. Ekonomou (Eds.): ISH 2019, LNEE 610, pp. 95–106, 2020.
https://doi.org/10.1007/978-3-030-37818-9_9

1 Introduction

Power systems all over the world are facing significant changes by different drivers that are changing the conditions of system operation. The significant increase of renewable resources is one of these drivers [1]. Due to their intimate nature, they induce more uncertainty to the grid, which needs more severe operation. The other game changer is the evolving of modern distribution networks. These systems become more complicated, regarding their new role in generation. Today, these grids can be known as key infrastructures both in consumption and generation because of the large presence of low and medium voltage connected distributed energy resources (DERs) and prosumers such as electric vehicles and energy storage systems (ESSs) [2].

On the one hand, in order to cope with these changes, distribution and transmission system operators (DSOs and TSOs) should incorporate different controllable equipment (CE) into the grid [3]. Generation units, FACTS devices, phase shifting transformers and ESSs are examples of these controllable resources. On the other hand, for guaranteeing the flexible operation of the grid, maintaining the system reliability and exploiting the efficient use of sustainable energy, the development of advanced monitoring solutions and control frameworks is necessary both for distribution and transmission grids [4].

In [5], an active/reactive power control strategy for ESS is proposed that takes into account the required profile of power for electrical grid by using local measurements. The authors of [6] propose a hysteresis-based PI-state control scheme for solving the problem of the limited fault ride-through capability of grid-connected DFIGs during faults and disturbances. They use local measurements for controlling the converter, which is connected to the ESS. The authors of [7] propose a novel parallel control for modular ESSs through the comparative analysis of various conventional parallel types of control by using local measurements for their proposed control scheme. Furthermore, in [8], the authors consider a distributed ESS-based control paradigm for enhancement of the power systems transient stability. They use local measurements (sensors) in a multi-agent control framework that includes both synchronous generators and ESSs. Kiaei and Lotfifard propose a model predictive control (MPC) method to improve the power systems transient stability by controlling the state of charge and discharge of ESS installed throughout power systems by using local measurements [9].

In [10], the authors use ESS for the transmission grid congestion management. In addition, they assess the effects of transmission congestion on the profitability of arbitrage by ESS in the electricity markets. Authors of [6] propose a scalable wide area coordinated control scheme by using distributed MPC on ESS to improve the rotor-angle stability in cases of large disturbances of the power system. Furthermore, they implement a local measurement system for the controllers of ESS in each area; however; the controllers are just able to perform coordinated control schemes. This coordinated control scheme is based on data exchange links.

The presented monitoring structures for control of ESS in previous works have just focused on a single control scheme; however; it is essential for grid scale ESSs to participate in different power services to make them financially feasible [11]. Therefore, this work takes advantage of wide area measurement system (WAMS) to present a configurable monitoring schemes for control of grid interactive ESS (GI-ESS). This will empower GI-ESS to participate in different power services in flexibility market, as well as, cover the drawbacks of decentralized, distributed and hierarchical control architectures.

In this paper, GI-ESS is referred to a grid scale ESS that is installed at primary substation. Therefore, it could be a representative for a single node in the well-known concept of active distribution networks [12]. Regarding the position of active substations as the interface with the transmission and distribution grids, they can participate in real-time operation enhancement of both girds. Overall, the main contributions of this paper are described as follows:

1. This works presents a monitoring scheme for the control of GI-BESS based on the synchrophasor measurements gathered by WAMS.
2. The provided situational awareness over the grid will enable the GI-BESS to participate in different power services in the flexibility market.
3. A basic power damping oscillation (POD) function is presented as an instance for power services that can be delivered by the GI-ESS.

The rest of this paper is organized as follows. Section 2 presents the main ideas behind the work and the proposed WAMS-based monitoring system for GI-BESS. The detailed design of the proposed monitoring system for POD function is illustrated in Sect. 3. The simulation results are presented in Sect. 4, and finally, Sect. 5 concludes the paper.

2 Synchrophasor Measurements Integration into Control

Monitoring systems are essential parts of every closed-loop control system. As it discussed in [11], many works can be found that take advantage of synchrophasor measurements into their control schemes; however; limited works can be found which have focused on the architecture of the monitoring system.

2.1 Synchrophasor Measurement Technologies

Synchrophasor Measurement Technologies (SMT) and their applications are encouraging solutions for the monitoring of emerging power system [13]. SMT can provide a wide range of real-time, online and offline applications such as stability assessment, congestion management and model validation for all monitoring, protection and control of the grid [14–16]. Generally, SMT includes phasor measurement units (PMUs), WAMS, and their applications in power systems monitoring and control.

PMU is an advanced technology in terms of measurements in power systems. This device is capable of gathering real-time measurement of voltage and current phasor (magnitude and angle), as well as, frequency and rate of change of frequency in a time synchronized manner. Precise time synchronized measurements characterized by GPS time tags and hybrid frequency and phase estimation algorithms are the major aspects that represent a substantial improvement in the concept of AC quantity measurement and provide power system operators many useful applications. Today, operation and planning of large power systems are enabled by the use of operation infrastructure, advanced measurement technology and information tools that WAMS provides. Typically, power system operators use WAMS as a standalone system. Therefore, these days, WAMS defined as a complimentary monitoring system beside SCADA, which is the conventional tool for power system operation and control.

For this purpose, the main function of WAMS is enhancing the real-time "situational awareness" of grid operator's that is essential for safe and reliable operation of the system. Furthermore, WAMS also supports post-event analysis for significant disturbances that occurs in the system. It is expected that WAMS and SMT become integrated into grids' real-time control centres (CC) in the near future [17].

Unlike conventional monitoring systems such as SCADA, energy management system (EMS) and substation monitoring system, SMT can be used in wide variety of control schemes including communication-based (centralized, decentralized and distributed) and autonomous (or local) control schemes, as well as, any hybrid combination of these schemes. Figure 1 shows different monitoring systems regarding their applications in control schemes and phenomena monitoring window. Regarding this figure, it could be note that:

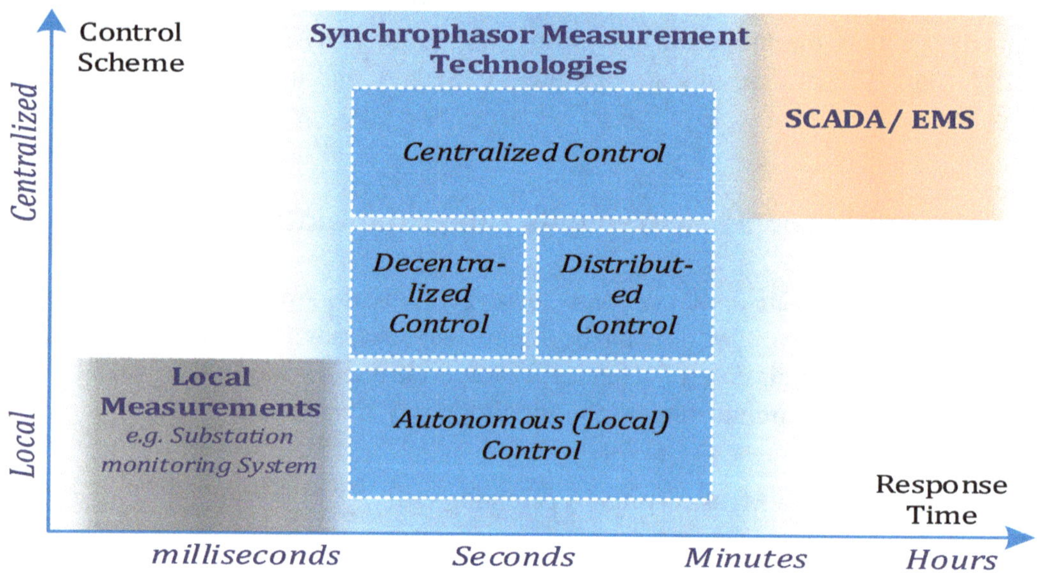

Fig. 1. Monitoring systems regarding time response and available control schemes.

- SMT can detect phenomena within the range of milliseconds to some minutes; however; there are modern PMUs that supports slow mode recording to detect phenomena with much lower occurrence frequency.
- Local measurements, such as substation monitoring systems are designed to provide measurements for local control loops. In addition, they can provide measurements for protective relays in substations and log the switching events, as well as, faults waveforms in the substation logging data centre.
- A typical RTU can gather and send electrical quantities such as voltage magnitude, active and reactive power in range of 1 to 5 s. The SCADA system receives and

archives these values. In addition, visualization part of SCADA system provides information about the status of the grid and its equipment for the operators in the CC.

- Generally, EMS uses the data gathered by SCADA system and smart meters for managing the energy production and consumption in the grid.

2.2 General Structure of SMT Integration into GI-ESS Control

In this paper, PMU and WAMS are taken into account as the local and wide area sources of synchrophasor measurements in concept of SMT.

The results of measurements-based algorithms such as oscillation detection, online transient stability assessment, online voltage stability assessment which can be implemented in local PDCs within power substations or CCs [18] are the sources of developed knowledge for SMT. This knowledge is extracted based on the synchrophasor measurements collected by WAMS or any other hybrid source of data. This integration provides situational awareness over the grid for the GI-ESS controller to enable it participation in dynamic stability improvement.

General structure is presented for SMT integration into control of GI-ESS in Fig. 2. This general structure shows how the local GI-ESS Controller can control the ESS Converter within the active substation, and how it can receive data from different monitoring systems to set the reference values for injecting/absorbing active and reactive power of the ESS.

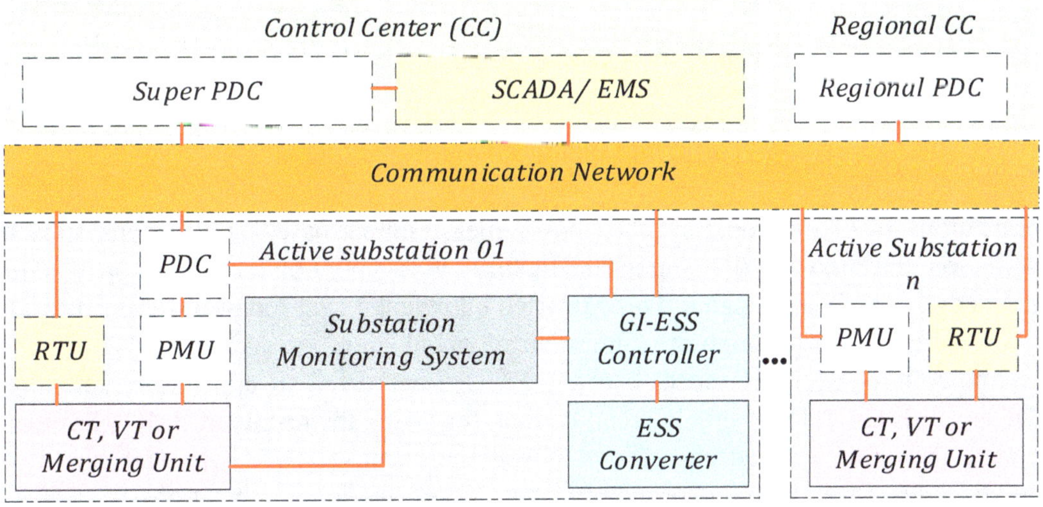

Fig. 2. General structure of SMT integration into GI-ESS control.

Regarding this general structure, different autonomous and communication-based control schemes can be implemented for the GI-ESS controller [12]. In this general structure, it is possible for GI-ESS to participate in inter-area interaction such as inter-area oscillation damping because of the available communication link between different Regional CCs.

2.3 Proposed WAMS-Based Solution for Monitoring System

Figure 3 shows the structure of WAMS-based solution for monitoring system. Regardless of the implemented WAMS architecture (hierarchical, centralized or hybrid), in the presented design, WAMS provides a historical database and a real-time stream of synchrophasor measurements of whole system.

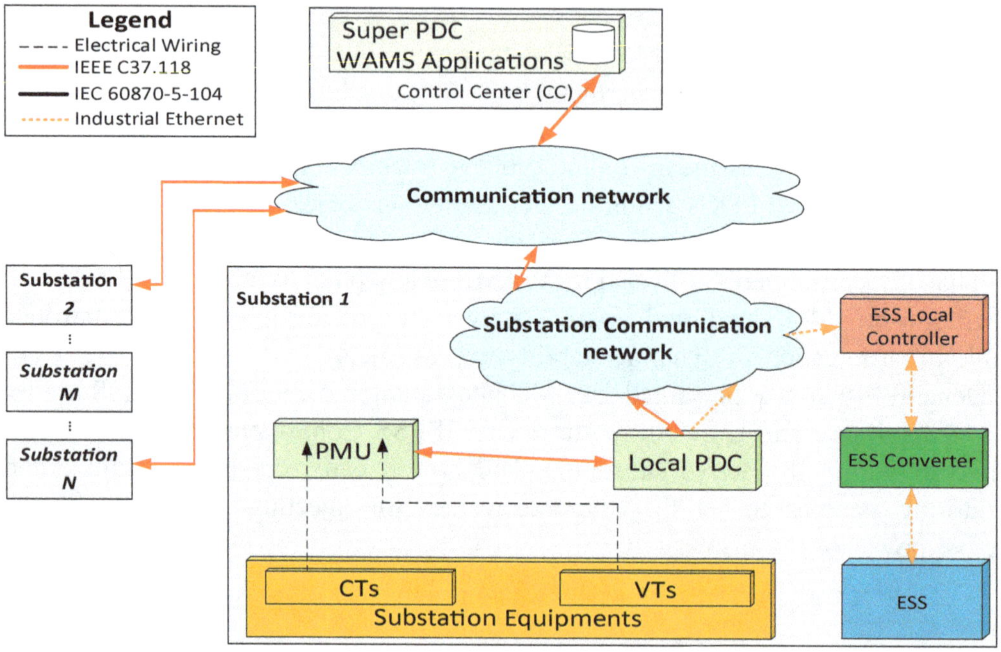

Fig. 3. WAMS-based solution for control.

In Fig. 3, CTs and VTs provide suitable signals for PMU, then, the PMU sends the synchrophasor measurements to the local PDC (if applicable) or Super PDC. Wide area applications in CC or local synchrophasor measurement-based applications such as oscillation detection and online transient stability assessment on local PDC provide the knowledge over the grid status. This extracted knowledge over the grid is transmitted to ESS local controller through the communication link.

The communication protocol between PMU, Local PDC (if applicable) and super PDC is IEEE C37.118. It should be noted that, regarding the results of PMU placement studies, PMUs are located on specific substations to make the voltage and current of specific lines observable, consequently, there could be some substations in the area (e.g. Substation M) without any installed PMUs.

ESS local controller can send the reference values for exchanging active and reactive power to the ESS converter regarding the grid and ESS status. For instance, the received knowledge over the grids identifies a local oscillation mode in the area of the GI-BESS, which enables the controller of GI-BESS to calculate the exact amount of active power, which is needed to be injected for damping of that specific mode. If the storage can provide that amount of active power, regarding the status of stored energy, the GI-BESS can participate in flexibility market for this power service.

One of the outmost aspect of this design is the need for implementation of real-time Ethernet protocol. The reason for this protocol selection is the need of deterministic communication link implementation in the closed-loop control systems. Because of this reason, it is essential to implement the real-time deterministic communication channel between each active substation and CC.

In addition, strict time stamping of PMU data can be used in control algorithms to suppress variations in communication channels for synchronization of local control loop to (global) WAMS measurement.

3 WAMS-Based Power Oscillation Damping

Low frequency inter-area oscillations can limit the power transfer between areas of power system [19]. Therefore, it is essential to increase the maximum power transfer (P_{Max}) between areas. Below, the structure of proposed controller for POD function is discussed.

Based on [20, 21], the objective of the Control Lyapunov Function (CLF) is to choose the value of a control variable in such a manner to make the output of the CLF non-positive. This is to ensure the stability of the system.

In this study, the WAMS delivers system-wide frequency and voltage angle measurements to the POD controller of the GI-BESS (δ_i, ω_i). These values are directly measured by phasor measurement units (PMUs) installed at generation substations of the system and sent via communication to the regional/super phasor data concentrator (PDC). Hence, it is essential to establish a communication link between PDC and the GI-BESS. For generation plants with more than one generation unit, an equivalent generation unit is assumed, therefore the *i*-index indicates the number of generation plant.

It is essential to note that in this study and for the sake of simplicity, all the communication links are assumed to be time deterministic without any possibility of data collision and delay.

The control diagram which is depicted in Fig. 4 shows the control scheme of a simple POD controller. The main idea behind this controller is taken from [21], which has implemented an POD for SVC. In [21], the generations of the system are divided into two critical and non-critical groups using coherency analysis, then, based on the principles of generalized one-machine infinite bus (GOMIB) system, the δ_{GOMIB} and ω_{GOMIB} are calculated for each sample of measured δ_i and ω_i. These calculations are done by wide area damping controller (WADC) as an application at the control center of WAMS.

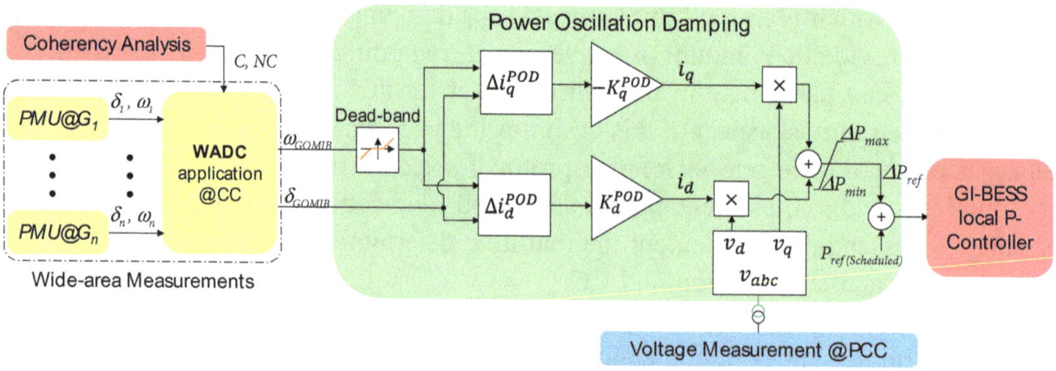

Fig. 4. Proposed wide area controller scheme for POD functionality

The POD block uses CLF as a representative for energy function of the system to mitigate the power oscillation by applying a change in the active power reference of the BESS Converter, ΔP_{ref}.

The single machine equivalent (SIME) method from [21] is used to obtain the proposed P-based control laws for the POD function, yielding

$$i_q = -k_q^{POD} \cdot cos(\delta_{GOMIB})\ \omega_{GOMIB} \tag{1}$$

$$i_d = k_d^{POD} \cdot sin(\delta_{GOMIB})\ \omega_{GOMIB} \tag{2}$$

where, i_q and i_d are q- and d-axis currents, k_d^{POD} and k_q^{POD} are time-varying tuneable gains that should be tuned optimally for different operational scenarios, and δ_{GOMIB} and ω_{GOMIB} are the angle and angular velocity of GOMIB. The ω_{GOMIB} and δ_{GOMIB} are calculated as

$$\omega_{GOMIB} = (\Sigma_{i\in C} m_i)^{-1} \Sigma_{i\in C} m_i \omega_i - (\Sigma_{i\in NC} m_i)^{-1} \Sigma_{j\in NC} m_j \omega_j \tag{3}$$

$$\delta_{GOMIB} = (\Sigma_{i\in C} m_i)^{-1} \Sigma_{i\in C} m_i \delta_i - (\Sigma_{i\in NC} m_i)^{-1} \Sigma_{j\in NC} \delta_j \omega_j \tag{4}$$

Where subscript C indicates the machines inside the critical group and subscript NC indicates the machines inside non-critical group in the concept of SIME.

Finally, the amount of reactive power to be exchanged by GI-BESS for POD ($P_{POD} = \Delta P_{ref}$) is calculated using $P_{POD} = v_q.i_q + v_d.i_d$, which is added to the P_{ref} which is scheduled for the GI-BESS.

4 Simulation and Results

To validate the proposed controller, the IEEE 39-bus system is used. The grid is implemented in DIgSILENT Power Factory and it is assumed this network comprises three principle areas, where a 200 MW/200 MWh GI-BESS with an initial state of charge (SoC) level of 50% is installed at *Bus 14*. The GI-BESS scheduled power is set to inject active power to the grid.

Regarding the installed capacity of this test system, the capacity of GI-BESS is almost 1% of total installed capacity (16800 MW) which is an acceptable range for a grid scale BESS.

In order to obtain good POD performance, it is necessary to find the optimal values for k_d^{POD} and k_q^{POD}. In this paper, a hybrid optimization based on combination of DE and L-BFGS-B optimization techniques is applied to find the optimal value for k_d^{POD} and k_q^{POD}.

As, the main focus of this paper is on the measurement system, the formulation of optimization problem is out of scope and will be discussed in future works. In addition, in this paper and for the sake of simplicity, it is assumed that $k_d^{POD} = k_q^{POD} = k$. The optimal value for k is 381.92.

It is essential to note that, the optimal value for k changes regarding the system's operation point. Therefore, it is recommended to update the gain value based on the grid status. However, in this paper and for an instance of power service for GI-BESS, just a single operation scenario is discussed.

In this operation scenario, the system undergoes a three-phase short-circuit event on line 14–15 at t = 1 s, which is cleared after 100 ms, causing an inter-area oscillation on the tie-lines 09–39 and 14–15.

As it depicted in Fig. 5, the FFT analysis of lines flow during the fault indicates the frequency of this oscillation. The modes shapes of this system after modal analysis for f = 0.645 Hz shows that two major groups of synchronous generators oscillate against each other.

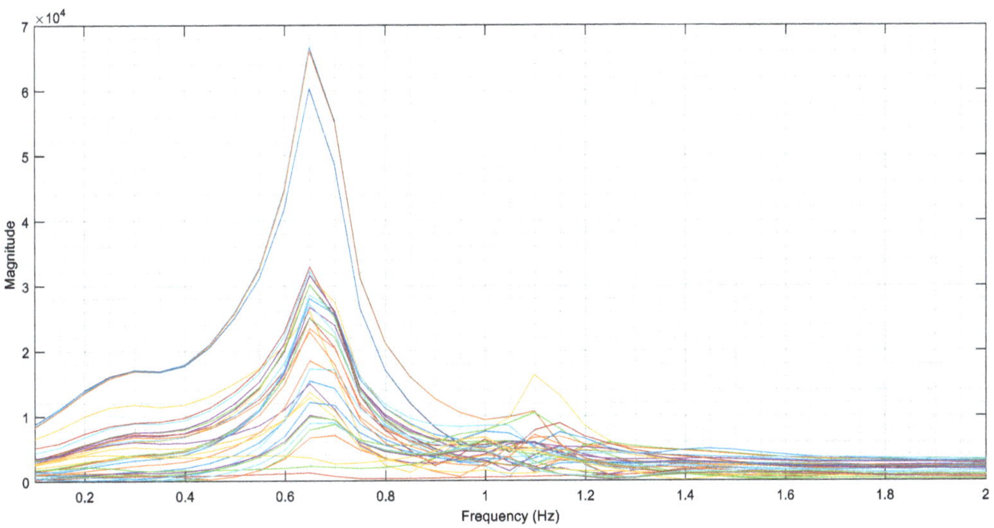

Fig. 5. FFT analysis of active power for all lines of IEEE 39 bus test case system.

Figure 6 depicts the active power of Line 14–15, for three different cases: without GI-BESS, with GI-BESS, and GI-BESS with the proposed controller. However, installing a GI-BESS improves the POD, this figure clearly shows the excellent

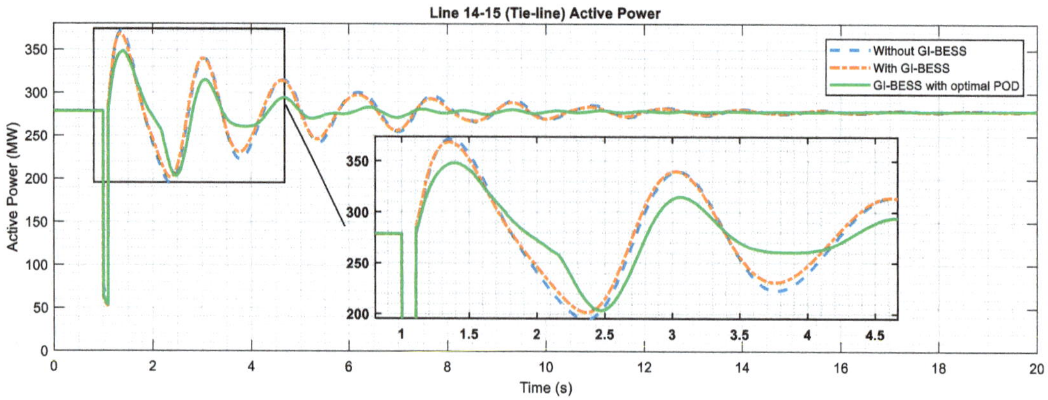

Fig. 6. The active power oscillation of Line 14–15, without GI-BESS and with GI-BESS with and without proposed POD controller.

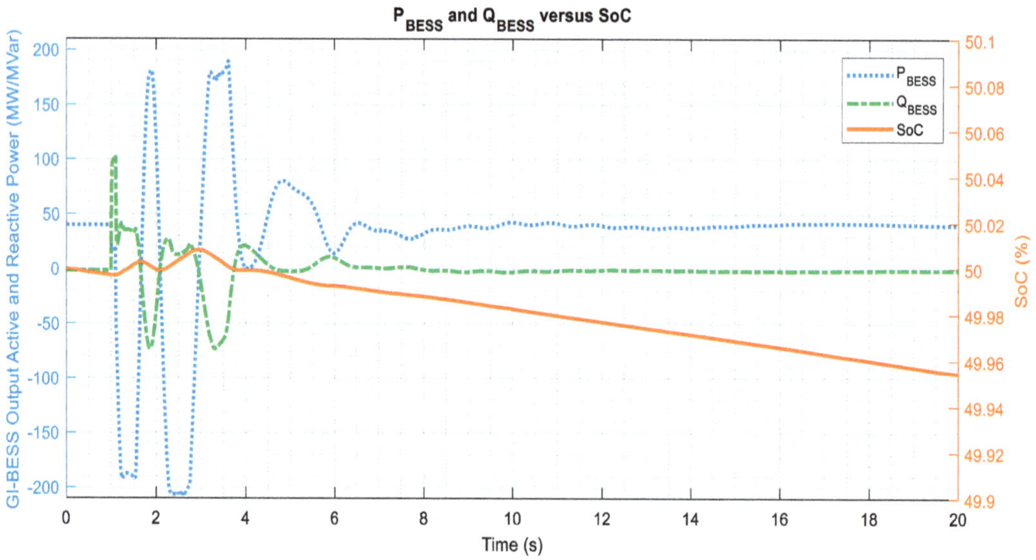

Fig. 7. SoC level versus the GI-BESS active and reactive power exchange.

performance of the proposed controller in regards to shaving the peak and increasing the damping ratio of the system.

For making GI-BESS available to participate in both energy and power services, it is important to assess the amount of stored energy, which could be injected to the grid and the available capacity for absorbing energy from the grid. The SoC can be defined as a good measure to find the effect of power oscillation damping enhancement as a power service on regular energy services of the GI-BESS. Therefore, to assess the effect of the proposed controller on the SoC level of the GI-BESS, a comparison is presented in Fig. 7.

Clearly, the impact of the proposed controller on the SoC level is negligible. This indicated that the GI-BESS was delivering the active power to the grid regarding the dispatch and participating in power oscillation damping simultaneously.

5 Conclusion

This paper presents a practical scheme for integration of SMT into the control of GI-ESS by delivering real-time situational awareness over the status of the grid. Furthermore, a basic CFL based POD function is implemented as an instance to assess this integration. This integration enables the GI-ESS to participate in both power and energy services of flexibility market, simultaneously. The simulation results prove the efficiency of the proposed monitoring system by showing the negligible impact the participation in power services has on the SoC of the GI-BESS, as well as, the excellent ability of GI-BESS to mitigate low frequency inter-area power oscillations.

References

1. Atanackovic, D., Clapauch, J.H., Dwernychuk, G., Gurney, J., Lee, H.: First steps to wide area control. IEEE Power Energy Mag. **6**(1), 61–68 (2008)
2. Pérez Arriaga, I., Knittel, C., et al: Utility of the future. In: An MIT Energy Initiative Response (2016)
3. Novosel, D., Madani, V., Bhargave, B., Vu, K., Cole, J.: Dawn of the grid synchronization. IEEE Power Energy Mag. **6**(1), 49–60 (2008)
4. Bevrani, H., Mitani, Y., Watanabe, M.: Power System Monitoring and Control. Wiley, Hoboken (2014)
5. Lawan, M.M.G., Raharijaona, J., Camara, M.B., Dakyo, B.: Power control for decentralized energy production system based on the renewable energies. In: ICRERA 2017, pp. 1132–1138 (2017)
6. Tourou, P., Chhor, J., Günther, K., Sourkounis, C.: Energy storage integration in DFIG-based wind energy conversion systems for improved fault ride-through capability. In: ICRERA 2017, pp. 374–377 (2017)
7. Ahn, J.H., Lee, B.K.: A novel parallel control for modular energy storage system achieving high performance, redundancy and applicability. In: ICRERA 2017, pp. 5–8 (2017)
8. Hammad, E., Farraj, A., Kundur, D.: on effective virtual inertia of storage-based distributed control for transient stability. IEEE Trans. Smart Grid **3053**(c), 1–10 (2017)
9. Kiaei, I., Lotfifard, S.: Tube-based model predictive control of energy storage systems for enhancing transient stability of power systems. IEEE Trans. Smart Grid **3053**(c), 1–10 (2017)
10. Wang, Y., Dvorkin, Y., Fernandez-Blanco, R., Xu, B., Kirschen, D.S.: Impact of local transmission congestion on energy storage arbitrage opportunities. In: IEEE Power Energy Society General Meeting (2018)
11. Kim, M., Kwasinski, A.: Decentralized hierarchical control of active power distribution nodes. IEEE Trans. Energy Convers. **29**(4), 934–943 (2014)
12. Torkzadeh, R., Eliassi, M., Mazidi, P., Rodriguez, P.: Synchrophasor measurements for control of grid interactive energy storage system: design alternatives for monitoring system. In: ICRERA 2018, Paris (2018)
13. Aminifar, F., Fotuhi-Firuzabad, M., Safdarian, A., Davoudi, A., Shahidehpour, M.: Synchrophasor measurement technology in power systems: panorama and state-of-the-art. IEEE Access **2**, 1607–1628 (2014)
14. Phadke, A.G., Thorp, J.S.: Synchronized Phasor Measurements and their Applications, vol. 1. Springer, Heidelberg (2017)

15. Ahmadzadeh-Shooshtari, B., Torkzadeh, R., Kordi, M., Marzooghi, H., Eghtedarnia, F.: SG parameters estimation based on synchrophasor data. IET Gener. Transm. Distrib **12**, 2958–2967 (2018)
16. Isazadeh, G., Kordi, M., Eghtedarnia, F., Torkzadeh, R.: A new wide area intelligent multi-agent islanding detection method for implementation in designed WAMPAC structure. Energy Procedia **141**, 443–453 (2017)
17. Hadley, M.D., McBride, J.B., Edgar, T.W., O'Neil, L.R., Johnson, J.D.: Securing wide area measurement systems. US Department Energy (2007)
18. He, M., Vittal, V., Zhang, J.: Online dynamic security assessment with missing PMU measurements: a data mining approach. IEEE Trans. Power Syst. **28**(2), 1969–1977 (2013)
19. Torkzadeh, R., Nasr-Azadani, H., Damaki-Aliabad, A.: A genetic algorithm optimized fuzzy logic controller for UPFC in order to damp of low frequency oscillations in power systems. In: ICEE 2014 (2014)
20. Latorre, H.F., Ghandhari, M., Söder, L.: Active and reactive power control of a VSC-HVDC. Electr. Power Syst. Res. **78**(10), 1756–1763 (2008)
21. Ghandhari, M.: Application of control Lyapunov functions to static var compensator. In: IEEE International Conference on Control Applications, pp. 1–6 (2002)

Icing Analysis of Kleče-Logatec Transmission Line with Two-Level Icing Model

Dávid Szabó[1(✉)], Bálint Németh[1], Gábor Göcsei[1], Viktor Lovrenčić[2], Nenad Gubeljak[3], Uršula Krisper[4], and Matej Kovač[5]

[1] Budapest University of Technology and Economics, Budapest, Hungary
szabo.david@vet.bme.hu
[2] C&G d.o.o., Ljubljana, Slovenia
[3] Faculty of Mechanical Engineering, University of Maribor, Maribor, Slovenia
[4] Elektro Ljubljana d.d., Ljubljana, Slovenia
[5] OTLM d.o.o., Ljubljana, Slovenia

Abstract. FLEXITRANTORE as a HORIZON 2020 project aims to develop an integrated platform for the next generation flexible electricity transmission system. During the project the investigation and implementation of state-of-the-art technologies will be realized, like storage systems, new market designs and business models, Dynamic Line Rating (DLR) technology or de-icing methodologies. The objective of FLEXITRANTORE's WP7 work package is to develop a DLR based anti-icing technology for transmission lines, therefore make the grid more reliable [1].

The aim of this paper is to present the investigation of complex icing system for transmission grid. The introduced investigation forms a two-level method, which in the first stage gives a warning for system operators, when the environmental factors are appropriate for ice formation on conductors. Then a mathematical model forecasts the radius and mass of the expected ice layer. The second level of the system monitoring the ice-formation on the phase conductors with line-monitoring sensors, thus the system operators get real information about the line condition. The operation of the methodology is also investigated, which are also presented through case studies.

Keywords: Flexitranstore · Overhead line · DLR · Ice prediction · Ice detection

1 Introduction

An Integrated Platform for Increased FLEXIbility in smart TRANSmission grids with STORage Entities and large penetration of Renewable Energy Sources (hereinafter referred to as FLEXITRANSTORE) is a project supported by the European Union's Horizon 2020 research and innovation programme. The main purpose of FLEXI-TRANSTORE is to improve the pan European transmission network to a high level of flexibility with high interconnection levels [1].

B. Németh and L. Ekonomou (Eds.): ISH 2019, LNEE 610, pp. 107–115, 2020.
https://doi.org/10.1007/978-3-030-37818-9_10

The aim of FLEXITRANSTORE's Work Package 7 is to develop a de-icing method for transmission lines based on Dynamic Line Rating (DLR) calculation process. DLR is a weather-based calculation methodology for uprating transmission capacity of existing power lines, thereby which can be used as lines with high temperature conductors. DLR offers temperature tracking calculation of the wires for the safety operation during extreme conditions, which can be reached as the result of prediction of local thermal overloads. On the other hand, ice layer can be formatted on the conductors in the winter extraordinary weather conditions, which should be also handled. A complex ice prediction model under development on the basis of DLR's environmental monitoring function. In the first stage, one of the demonstration lines was investigated based on the collected field data, which gives the principle of the method [2, 3].

2 BME's Ice Prediction Model

Besides the geometry of the conductor, local environmental conditions, such as rainfall, ambient temperature, humidity, wind speed and direction, also play an important role in the formation of ice layer on the surface of the conductors. These parameters determine the structural properties of the resulting ice layer and thus its properties. Based on these environmental factors, three types of ice can be distinguished, which can cause high mechanical extra load to the conductors through high-adhesion and density. These three ice types are wet snow, glaze and hard rime.

BME's ice type determining system is established to predict the expected ice type based on environmental parameters, on which based the ice layer diameter and extent of extra mechanical load can be calculated according to the actual ice type. The algorithm takes into account the ambient temperature, precipitation type and intensity, relative humidity and also the temperature of the conductors in order to determine the expected ice type. The results of the system can be the following: wet snow, mixture of wet snow and glaze, glaze, mixture of glaze and hard rime, hard rime or ice formation is not expected. Ice can only shape when conductor temperature below 2 °C, but due to the uncertainty of the conductor temperature calculation model and the deviation of line monitoring devices, this threshold value was set to 3 °C in the model, which appears as a safety factor while it can be also increase the number of false alarms [4].

The structure of the ice layer deposited on the overhead line conductors largely depends on the type of precipitation, which through several parameters - water droplet/snowflake velocity and mass concentration, collision efficiency, adhesion factor, deposition factor - influences the forming ice layer. In this way, the ice layer will be accreted differently for different types of ice, so the calculation of the thickness of the resulting ice sleeve and the consideration of the extra mechanical load caused by it should be calculated in different ways depending on the type of ice. BME's ice determining system use Lacavalla et al. model [5, 6] for wet snow calculation, Pytlak et al. model [7] for glaze computation and Shao et al. model [8] for hard rime estimation.

3 Ice Detection System of OTLM

3.1 Development of OTLM Line Monitoring Sensor

Severe ice storm was observed in Slovenia in 2014, which indicates the demand for new safety measures during overhead lines operations. For this purpose, one of the additional safety precautions was to inspect the overhead lines when received information from national weather service that ice storm is possible in the given region. Based on this forecast information the TSO should check if ice storm is affected the overhead line or not. Unfortunately, some overhead lines are located high in the hills and approach is nearly impossible in case of snowy winter time. These facts encourage the manufacturer to build a camera into OTLM line monitoring sensor, which can be used for monitoring overhead lines and to check the ice status on overhead line conductors.

Fig. 1. OTLM line monitoring sensor

Beside the presence of the ice, operator can check what kind of ice is present - glaze ice, wet snow, etc. In case of icing event system operator is able to heat the conductor, then the critical point will be the tower. Thus, OTLM's camera was turned towards the tower to check the conductor and the tower at once. This feature enables the operator to check the status of overhead lines in real-time without the presence of maintenance personnel and act accordingly on the real status.

3.2 OTLM's Ice Detection Function

This chapter presents the concept of the application and the relation between the geometry and load parameters on the catenary curve when ice or heavy snow builds up.

Mathematical model has been developed for sag and horizontal force calculation. The model includes installation conditions and conductor characteristics and determines the interdependence between conductor sag and horizontal force for actual conductor temperatures based on mechanical and physical characteristics of the conductor, conductor weight and sag size. Combining measurements of conductor geometry and sag at several conductor temperatures with software are used for calibration of the sag and angle function. Ensuring conformity is crucial for the implementation of the ICE-ALARM function, since a continued growth of discrepancy between the measured and calculated angle in ambient conditions is a sign of glaze ice on the conductor.

The parameters of the catenary curve at the temperature of the freezing rain represent the initial state of the activation of the ICE-ALARM computer algorithm. If favorable conditions for the formation of ice appear during the continuous monitoring of the conductor condition and condition on the route in the surroundings of the meteorological station, then it is possible to estimate the amount of additional loading and the ice thickness on the basis of the change in the angle of inclination and by knowing the tension-deformation behavior of the conductor at increased loading.

Figure 2 shows the change in the angle in accordance with the model and the angle measured by the inclinometer. White circles present actual average angles as a function of average temperature of conductor measured in the time interval of 30 s. Red circles presents the expected behavior of the conductor and/or a change in the angle due to the

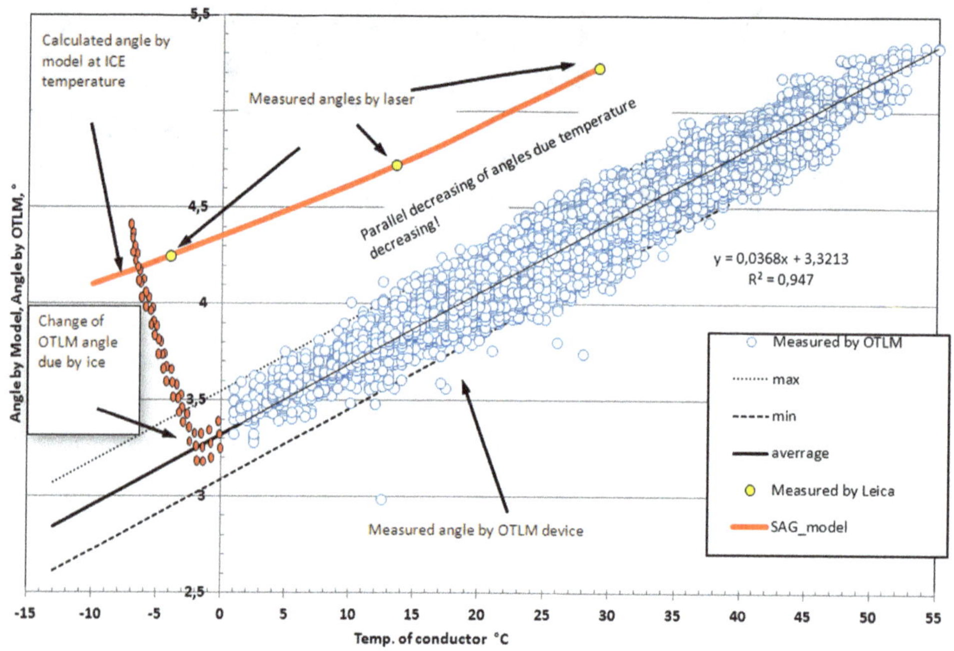

Fig. 2. Change in an angle at the OTLM device position depending on temperature

build-up of the ice on the conductor. The continuous red line represents the angle of inclination depending on temperature according to the mathematical model. If an angle significantly increases in the meteorologically favorable ice conditions and the temperature inversion and if the calculated angle significantly differs from the angle measured by inclinometer, the application informs the operator that ice has built up on the conductor [9–11].

4 Two-Level Icing Model

Based on BME's ice prediction model and OTLM's ice detection function a two-level model was established and under implementation for Kleče -Logatec transmission line. In this way the ice accretion can be predicted based on weather forecast according to BME's ice type determining system co-operated with national weather service (ARSO). The accreting ice layer can pose a threat if the sleeve radius exceeds 10 mm or if the extra mechanical load caused by it exceeds 1 kg/m during the icing event. Therefore, the threshold values for ice alarm settled to be according to these limitations.

Furthermore, if ice alarm was sent to the system operators, they can monitor the real field conditions of the conductors with OTLM device, which can not only measure the inclination of the monitored span, but also the conductors state can be observed with the built-in camera. Therefore, the ice prediction model's results can be compared with the actual conditions, and the required intervention can be determined according to the growth rate of the ice layer.

5 Case Studies

To illustrate the operation of the two-level icing model, some case studies are presented here. Although, there was a "green winter", which means there was no considerable icing, only some snowing events occurred, nevertheless the operation of the model can be showed through these snowing events. Case studies was made for Kleče-Logatec 110 kV single circuit transmission line equipped with 240/40 mm^2 ACSR conductors.

5.1 20 November 2018

BME's model predicted wet snow and glaze ice types based on weather forecast for different grid points. The expected ice thickness was 5 to 6 mm for glaze and 10 to 14 mm for wet snow. Figure 3 shows the accretion of glaze ice depending on precipitation intensity.

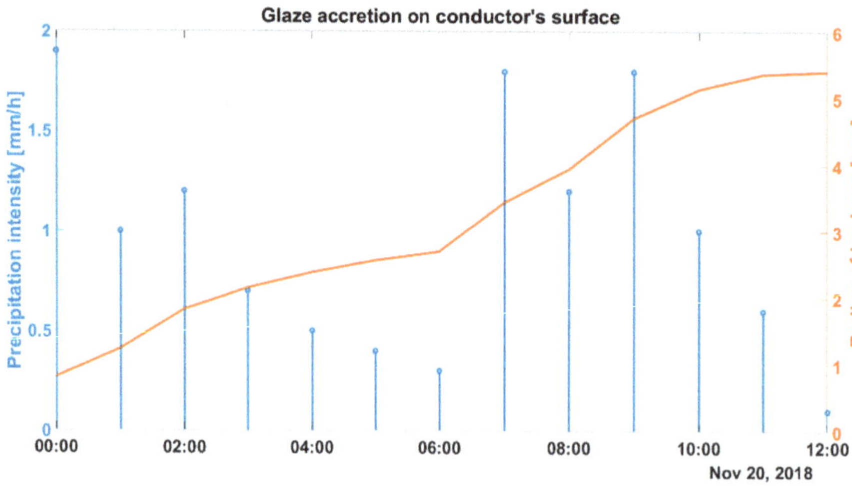

Fig. 3. Glaze accretion according BME's model - 20 November 2018

On the other hand, as Fig. 4 shows the image captured by OTLM device, there is a slight ice layer can be seen on the bottom of the wire.

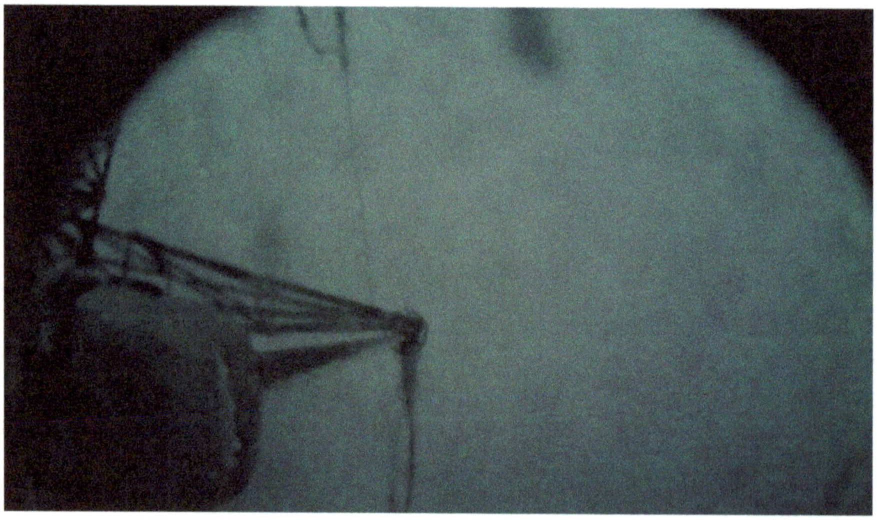

Fig. 4. Real state of a phase conductor - 20 November 2018

5.2 18 January 2019

A mixed type of ice from wet snow and glaze was anticipated according to BME's ice prediction model with a thickness between 9 to 12 mm for the different forecast grid points. The expected ice formation is shown in Fig. 5.

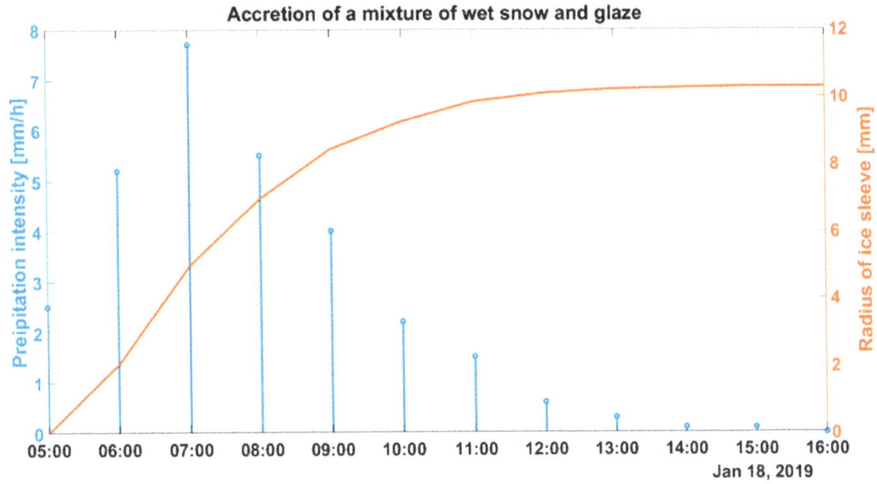

Fig. 5. Glaze accretion according BME's model - 18 January 2019

The real field conditions are shown in Fig. 6, where a huge snow deposit can be seen front of the camera, and a layer of ice on the tower.

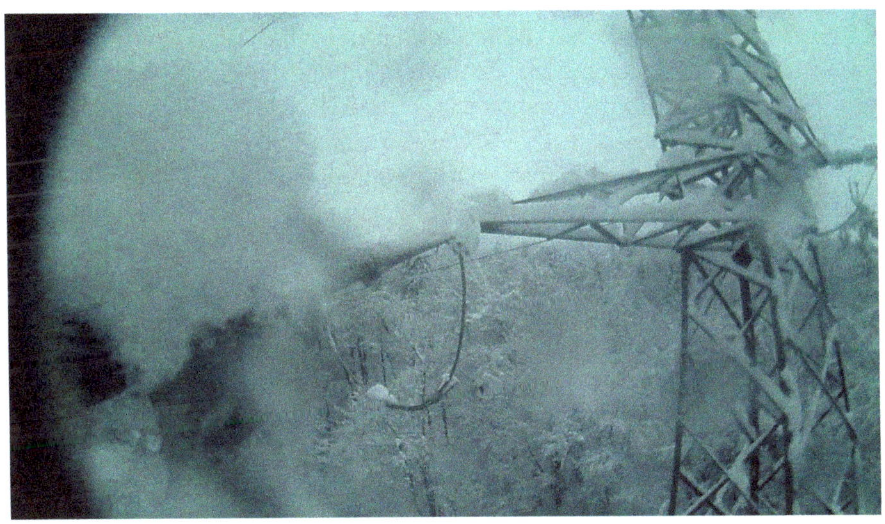

Fig. 6. Real state of a phase conductor - 18 January 2019

5.3 Summary of Case Studies

The operation of the model was investigated in 2018–19 winter time, when only slight ice formed on the conductors mostly form wet snow. BME's ice prediction model forecasted properly the ice formation, while the quantitative estimation should be fine-tuned, when significant ice sleeves will occur. On the other hand, OTLM device offers an appropriate solution for real-time monitoring of the conductors, which can be the basis to the intervention for system operators.

6 Conclusion

According to FLEXITRANSTORE's project this paper presented the development of a two-level ice prediction and detection model for high voltage overhead lines. The first level is a weather-based system, which aims to predict the possibility of different ice types – wet snow, glaze, hard rime – accretion on conductors. The model is able to calculate the radius of the ice sleeve and its mechanical extra load based on the accreting ice layer's type. On the second level a computer algorithm was developed for re-calculation of the sag and tensile strains in the conductor. It takes into account the actually measured form of the catenary curve of the conductor on the presented span at the conductor temperature measured by OTLM. Based on the knowledge about the change in the sag of the catenary curve and the tensile forces dependence on the temperature of the conductor and monitored weather conditions, it is possible to determine the moment of activation the ICE-ALARM application. Furthermore, OTLM sensor is able to monitor the actual state of the conductors with its camera.

The essence of the two-level system is the prediction opportunity combined with the real-time monitoring function. System operators get a forecast of the seriousness of the icing event in this way, while the intervention can be made according to the danger factor, therefore the number of unnecessary interventions can be reduced.

Acknowledgements. This work has been developed in the High Voltage Laboratory of Budapest University of Technology and Economics within the boundaries of FLEXITRANSTORE project, which is an international project. FLEXITRANSTORE (An Integrated Platform for Increased FLEXIbility in smart TRANSmission grids with STORage Entities and large penetration of Renewable Energy Sources) aims to contribute to the evolution towards a pan-European transmission network with high flexibility and high interconnection levels.

References

1. FLEXITRANSTORE: European Union's Horizon 2020 Project. http://www.flexitranstore.eu/
2. Rácz, L., Szabó, D., Göcsei, G., Németh, B.: Grid management technology for the integration of renewable energy sources into the transmission system. In: 2018 7th International Conference on Renewable Energy Research and Applications (ICRERA), Paris, 2018, pp. 612–617 (2018)
3. Halász, B.G., Németh, B., Rácz, L., Szabó, D., Göcsei, G.: Monitoring of actual thermal condition of high voltage overhead lines. In: Camarinha-Matos, L., Adu-Kankam, K., Julashokri, M. (eds.) Technological Innovation for Resilient Systems. DoCEIS 2018. IFIP Advances in Information and Communication Technology, vol. 521. Springer, Cham (2018)

4. Szabó, D.: Development of dynamic line rating system. MSc thesis, Budapest University of Technology and Economics, Budapest (2019)
5. Lacavalla, M., Marcacci, P., Frigerio, A.: Forecasting and monitoring wet-snow sleeve on overhead power lines in Italy, pp. 78–83 (2015). https://doi.org/10.1109/eesms.2015.7175856
6. Lacavalla, M., Bonelli, P., Mariani, G., Marcacci, P., Stella, G.: The WOLF system: forecasting wet-snow loads on power lines in Italy. In: 14th International Workshop on Atmospheric Icing of Structure, 8–13 May 2011, Chongqing (2011)
7. Pytlak, P., Musilek, P., Lozowski, E., Arnold, D.: Evolutionary optimization of an ice accretion forecasting system. Mon. Weather Rev. **138**(7), 2913 (2010)
8. Shao, J., Laux, S.J., Trainor, B.J., Pettifer, R.E.W.: Nowcasts of temperature and ice on overhead railway transmission wires. Meteorol. Appl. **10**, 123–133 (2003). https://doi.org/10.1017/s1350482703002044
9. Lovrenčič, V., Gubeljak, N., Banić, B., Ivec, A., Kozjek, D., Jarc, M.: On-line monitoring for direct determination of horizontal forces vs. temperature and incline angle of transmission conductor. In: Fourth Session of Cired Croatian National Committee), 11th Symposium on power system management HRO-CIGRE, Opatija, Croatia, November 2014
10. Lovrenčič, V., Gabrovšek, M., Kovač, M., Gubeljak, N., Šojat, Z., Klobas, Z.: The contribution of conductor temperature and sag monitoring to increased ampacities of overhead lines (OHLs). In: DEMSEE 2015 10th International Conference on Deregulated Electricity Market Issues in South Eastern Europe, Budapest, Hungary, September 2015
11. Gubeljak, N., Lovrenčić, V., Banić, B., Pinterič, M., Jarc, M.: Application of an ice-alarm in the OTLM system. In: First South-East European CIGRE Conference, Portorož (2016)

www.ingramcontent.com/pod-product-compliance
Lightning Source LLC
LaVergne TN
LVHW082239200926
848484LV00008B/2375
* 9 7 8 1 0 1 3 2 7 0 7 7 2 *